The Baofeng Radio Bible

Your Essential Guide for Emergency Communication

Master Setup, Programming, and Operation for Effective Use in Everyday and Crisis Situations

Eliot Vance

Contents

Introduction

Chapter 1: A Brief History of Baofeng and Its Popularity Among Ham Operators and Preppers

The Origins of Baofeng

Baofeng, a Chinese electronics manufacturer, began making its mark in the world of communication equipment in the early 2000s. Though initially unknown outside China, the company quickly gained attention in the amateur radio world with its affordable and versatile two-way radios, particularly the UV-5R model, which launched in 2012. The UV-5R became Baofeng's flagship product and is widely recognized as a game-changer for both ham operators and preppers due to its blend of performance, features, and price.

The company's focus was on producing radios that were compact and user-friendly, with enough advanced features to appeal to enthusiasts. While many established brands such as Kenwood, Yaesu, and Icom dominated the amateur radio market, their products were often priced higher, making it difficult for beginners or casual users to get started. Baofeng filled this gap by offering functional radios that were accessible to everyone, democratizing two-way communication.

Affordability: The Key to Success

One of the most appealing aspects of Baofeng radios, especially for American ham operators and preppers, has been the price point. A typical Baofeng UV-5R costs around $30 to $50, while similar radios from competitors often range from $100 to $300. This made Baofeng an instant hit among budget-conscious hobbyists, beginners, and those looking to add multiple radios to their gear without breaking the bank.

Despite the low cost, Baofeng radios come with a robust set of features. The UV-5R, for example, is a dual-band radio capable of operating on both VHF (136-174 MHz) and UHF (400-520 MHz) frequencies. It includes features such as dual-watch mode, emergency alert functionality, and FM radio reception—all of which make it not just a functional communication tool, but a versatile one that appeals to both casual users and serious hobbyists.

Why Baofeng Is Popular Among Ham Radio Operators

Amateur (or ham) radio operators are individuals who use two-way radios for non-commercial communication, often as a hobby, but also for emergency preparedness and public service. In the United States, ham operators need to pass a licensing exam from the Federal

Communications Commission (FCC), which allows them to legally transmit on a wide range of frequencies not available to the general public.

For ham radio operators, Baofeng radios offer several benefits:

1. **Ease of Use**: Although Baofeng radios have a slight learning curve when it comes to manual programming, they are relatively easy to use for basic communication once set up. Ham operators can easily switch between frequencies, set privacy tones, and communicate with others in their network.
2. **Programmability**: Baofeng radios can be programmed using software like CHIRP, making it simple to set up multiple channels and frequencies, including local ham repeaters, public safety channels, and even weather alerts. This allows operators to have all necessary frequencies pre-programmed and ready to go.
3. **Durability**: Despite their low price, Baofeng radios are surprisingly durable. They can withstand the daily wear and tear of outdoor use, which is important for ham operators who often participate in field activities such as portable operation (also known as "pota" or "parks on the air") and emergency communication drills.
4. **Expandability**: Baofeng radios are highly customizable, with the ability to swap out antennas for better range, add external microphones for ease of use, and even upgrade batteries for longer operation. For ham operators who like to modify and optimize their gear, Baofeng offers plenty of flexibility.

Baofeng's Role in Emergency Preparedness and Prepping

Beyond the ham radio community, Baofeng radios have found a devoted following among preppers—individuals and groups who prepare for potential disasters, emergencies, or societal collapse. Preppers are keenly aware of the fragility of modern communication systems, which can be disrupted by natural disasters, power outages, or civil unrest. For these individuals, having a reliable method of communication that doesn't depend on the internet, phone lines, or infrastructure is critical.

Baofeng radios, especially the UV-5R model, have become a go-to tool for preppers for several reasons:

1. **Emergency Communication**: In a grid-down scenario, preppers can use Baofeng radios to communicate with family, neighbors, or other members of their preparedness community. The ability to broadcast on both UHF and VHF frequencies means they can find available channels for local communication, especially when cell towers or internet connections are down.
2. **Interoperability with Other Radios**: Baofeng radios can easily communicate with other handheld transceivers (HTs), as well as mobile and base station radios, making them an ideal tool for maintaining a communications network. Preppers can also tune

into local repeaters, weather alerts (NOAA), and emergency broadcasts, which are critical for staying informed during a disaster.

3. **Affordability and Accessibility**: Preppers often plan for worst-case scenarios, which means they need backup equipment. The affordability of Baofeng radios allows them to stock multiple units without spending excessive amounts of money. Having extra radios means they can be distributed among family members or neighbors in an emergency.

4. **Simplicity and Portability**: Baofeng radios are lightweight and portable, making them ideal for bug-out bags or emergency kits. In the event of a sudden evacuation, preppers can easily carry their radios, ensuring they have a means of communication no matter where they go.

The Role of Baofeng in American Disaster Preparedness

In the United States, natural disasters such as hurricanes, tornadoes, wildfires, and earthquakes can wipe out traditional communication infrastructure. During these times, amateur radio operators often step in to provide critical communication services to emergency responders and local communities. Many preppers have adopted Baofeng radios specifically for this purpose.

For example, during Hurricane Katrina in 2005, ham radio operators played a vital role in relaying information to rescue teams and providing a lifeline for stranded individuals. Since Baofeng radios can access amateur radio frequencies, they are particularly valuable for emergency communication when modern systems fail.

Additionally, the National Oceanic and Atmospheric Administration (NOAA) broadcasts weather alerts on specific frequencies, which can be accessed via Baofeng radios. Preppers and survivalists often pre-program these frequencies into their radios to receive real-time updates during emergencies, ensuring they stay informed about approaching storms or other hazards.

The Growing Popularity of Baofeng Radios in the Prepper Community

Baofeng radios have continued to grow in popularity within the prepper community, especially with the rise of online forums, YouTube channels, and social media groups dedicated to preparedness. Many of these platforms offer tutorials on how to program, use, and optimize Baofeng radios for specific scenarios, which has only fueled the brand's reputation.

Preppers often share lists of recommended frequencies, including local emergency services, prepper networks, and commonly used channels for group communication. Many preppers have also embraced the idea of having a "communications plan" that involves Baofeng radios as a primary or backup method of communication.

For many, owning a Baofeng radio has become synonymous with being prepared for the unexpected. Whether it's a natural disaster, a grid failure, or a societal collapse, Baofeng radios provide a reliable, affordable, and flexible solution for communication in times of crisis.

Conclusion

Baofeng's rise to prominence can be attributed to its affordability, versatility, and widespread appeal among ham operators and preppers alike. For ham radio enthusiasts, Baofeng offers a simple, accessible entry point into the world of amateur radio. For preppers, it provides a lifeline in emergencies, offering a reliable way to stay connected when traditional communication systems fail.

In a world where communication is essential, but often taken for granted, Baofeng radios serve as a critical tool for those looking to stay informed, prepared, and ready for any challenge. Whether you're exploring the world of ham radio or preparing for the unknown, Baofeng radios have earned their place as a trusted companion in the hands of enthusiasts and preppers across America.

Chapter 2: Why Baofeng Radios Are Essential for Communication in Emergencies

In an increasingly connected world, it's easy to assume that cell phones, the internet, and other modern technologies will always be available when you need them. But during times of crisis—whether it's a natural disaster, infrastructure failure, or a large-scale emergency—these systems can become unreliable or completely unusable. This is where the Baofeng radio shines, providing a lifeline for communication when other technologies fail.

Baofeng Radios: An Independent, Reliable Communication Tool

Baofeng radios are essential for emergency preparedness because they offer a level of independence that other devices do not. Unlike cell phones or internet-based communication systems, Baofeng radios rely on radio waves rather than infrastructure like cell towers or cables. This independence makes them a critical tool for staying in contact with others when traditional services are unavailable.

1. **No Reliance on Infrastructure**
   During a disaster, cell towers can be knocked out, phone lines can go down, and even satellite communication may be disrupted. Baofeng radios bypass these issues by operating on VHF (Very High Frequency) and UHF (Ultra High Frequency) radio bands, which do not require external infrastructure. As long as you have power in your radio and a line of sight or close proximity to the person or group you want to communicate with, you're still connected.
2. **Simple and Efficient Communication**
   In emergency scenarios, communication needs to be quick and effective. Baofeng radios allow you to communicate instantly at the push of a button. Whether you're coordinating rescue efforts, alerting others to hazards, or keeping family members in touch, the simplicity of two-way radio communication can be a lifesaver.
3. **Wide Range of Frequencies**
   Baofeng radios can access a broad spectrum of frequencies, making them highly versatile. They can be used to tune into local amateur (ham) radio networks, GMRS (General Mobile Radio Service), FRS (Family Radio Service), and even NOAA weather stations for real-time updates on hazardous conditions. The ability to switch between frequencies allows you to tap into local emergency services, find community communication channels, or simply talk to your group without interference.
4. **Range Flexibility**
   Baofeng radios are adaptable in their range. While they typically operate over shorter distances (a few miles, depending on the terrain and environmental factors), they can significantly extend their range when used with repeaters—specialized stations that

rebroadcast radio signals. This flexibility is especially important in rural areas, where cell coverage is sparse or nonexistent.

5. **Emergency Alert Features**
Many Baofeng models come equipped with emergency alert functions. With the press of a button, you can send out an emergency broadcast that alerts others to your situation. This feature is invaluable in life-threatening scenarios where you need to draw attention to yourself quickly.

6. **Portability and Power Options**
Baofeng radios are small, lightweight, and easily portable, making them perfect for survival kits or everyday carry. Additionally, they offer multiple power options. Baofeng radios typically come with rechargeable lithium-ion batteries, but they can also be powered by external battery packs or car adapters. In a long-term emergency, this flexibility ensures that your radio remains operational even if the power grid fails.

Communication in a Disaster: Real-World Applications

There are countless real-world examples where Baofeng radios can be a crucial asset:

- **Natural Disasters**: Whether you're dealing with a hurricane, earthquake, or wildfire, communication becomes vital as power grids go down and mobile networks are overwhelmed. Baofeng radios allow you to coordinate with neighbors, first responders, or local emergency organizations.
- **Grid Failures**: In a prolonged blackout or grid-down scenario, where utilities and communication lines are compromised, Baofeng radios provide a way to stay informed and connected. You can use them to listen in on emergency broadcasts or stay in touch with your preparedness group.
- **Outdoor Expeditions**: Whether you're hiking in remote areas or on a camping trip, Baofeng radios are an ideal backup in case you're out of cell range. In the event of an emergency, these radios can connect you with rescue teams or other hikers in the area.
- **Civil Unrest**: In situations of societal breakdown or civil unrest, having a reliable communication system that doesn't depend on commercial networks is essential. Baofeng radios offer secure communication for your group while keeping you informed of unfolding events.

How This Book Will Help You Master Your Baofeng Radio

This book is designed to walk you step-by-step through everything you need to know about your Baofeng radio. Whether you're a beginner who's never used a two-way radio before or a more experienced user looking to optimize your setup, this guide will provide the knowledge and practical skills to help you feel confident in using your radio, especially in emergency situations. Here's how this book is structured to make you a master of Baofeng radios:

1. **Understanding the Basics**
   We'll start with the fundamentals of radio communication, explaining key concepts such as VHF vs. UHF, radio frequencies, and how two-way radios operate. These core principles will help you understand the unique strengths of Baofeng radios and how they fit into the larger world of communication technology.

2. **Mastering Programming**
   One of the most critical skills for any Baofeng user is learning how to program the radio. This guide will take you through manual programming, showing you how to input specific frequencies and set up essential channels. Additionally, for those looking to streamline the process, we'll cover how to use CHIRP software for faster and more complex programming.

3. **Legal and Regulatory Knowledge**
   Using a Baofeng radio in the U.S. requires an understanding of Federal Communications Commission (FCC) regulations. We'll explain which frequencies require a license (such as ham and GMRS), and which are available for unlicensed use (FRS, MURS). This section will also clarify the steps to get your amateur radio license, ensuring you remain compliant with U.S. laws.

4. **Maximizing Range and Performance**
   While Baofeng radios are powerful out of the box, there are several ways to enhance their performance. This guide will explain how to boost your range using external antennas, optimize your battery life, and adjust your settings for maximum clarity and signal strength. For those operating in challenging environments (mountains, forests, urban settings), we'll also provide strategies to overcome common obstacles.

5. **Emergency Use Cases**
   A dedicated section of the book will focus on real-life scenarios where Baofeng radios can save lives. Whether you're preparing for natural disasters, power outages, or survival situations, we'll show you how to configure your radio for these events, including how to find local emergency channels and join prepper communication networks. In this section, you'll also learn how to set up a family or group communication plan.

6. **Practical Exercises**
   Learning about radios is one thing—using them effectively in practice is another. Throughout the book, you'll find exercises and real-world applications to help you practice what you've learned. This hands-on approach will build your confidence in using your radio, from basic operations like tuning into a local frequency to advanced techniques like repeater use and dual-watch monitoring.

7. **Maintenance and Troubleshooting**
   To ensure your radio is always in top working condition, this book will guide you through regular maintenance tips and solutions for common problems. We'll cover

everything from protecting your radio from water damage to resetting it if something goes wrong, ensuring that you'll always be prepared for unexpected situations.

8. **Resources for the Future**

    Finally, this guide will include resources for expanding your knowledge even further. From online communities and ham radio clubs to additional software and accessories, we'll point you to valuable tools and networks that can help you stay up to date with the latest in radio technology.

Conclusion

Mastering your Baofeng radio is about more than just understanding how it works—it's about being prepared for the unexpected. Whether you're planning for emergencies, venturing into the great outdoors, or simply expanding your communication skills, this book will equip you with the knowledge and practical tools to use your radio effectively.

By the time you finish reading this guide, you'll not only know how to program and operate your Baofeng radio, but you'll also have the confidence to use it when it matters most. From staying informed in a crisis to keeping in touch with your loved ones, your Baofeng radio will become a crucial part of your emergency toolkit—and this book will ensure you know exactly how to use it to its full potential.

Chapter 3: Understanding Radio Basics

Before diving deep into using your Baofeng radio, it's important to have a solid grasp of the basics of radio communication. This chapter will walk you through the fundamental principles of how radios work, key terms you'll encounter, and why Baofeng radios, like the popular UV-5R, have become a go-to choice for many radio enthusiasts and preppers alike.

What is a Two-Way Radio?

A two-way radio is a communication device that allows users to send and receive voice transmissions over specific frequencies. Unlike one-way radios, such as broadcast radios where you can only listen (think of FM or AM stations), two-way radios enable a user to both talk and listen, allowing for direct interaction. These radios are often referred to as "transceivers" because they both transmit and receive signals.

Two-way radios are widely used by a variety of people—emergency responders, hikers, truckers, ham radio operators, and preppers—because they enable instant, reliable communication without relying on modern infrastructure such as cell towers or the internet.

Basic Explanation of How Radios Work

At their core, radios work by converting sound (your voice) into electromagnetic waves that travel through the air. These waves are picked up by another radio receiver, which converts the waves back into sound. Here's a simplified breakdown of the process:

1. **Transmission**: When you speak into your two-way radio, your voice is converted into electrical signals. These signals are then modulated onto radio waves.
2. **Propagation**: The radio waves travel through the air over specific frequencies. The distance they can travel depends on the power of the radio, terrain, and whether repeaters (devices that boost signals) are used.
3. **Reception**: Another two-way radio tuned to the same frequency picks up the radio waves. The radio then demodulates the waves and converts them back into sound, allowing the person on the other end to hear your voice.

Unlike cellular or internet-based communications, which require complex networks and infrastructure, two-way radios use direct transmission from one device to another, making them highly reliable in situations where other communication methods might fail.

Differences Between Analog and Digital Radios

Baofeng radios, such as the UV-5R, are analog radios. But what's the difference between analog and digital radios?

1. **Analog Radios**:
   Analog radios transmit voice signals in a continuous wave, much like old landline phones or FM radio stations. The sound is transmitted directly as an electromagnetic wave, which means the quality of the audio can degrade over long distances or if there's interference. Analog radios, like those from Baofeng, are simpler to operate and widely used in various amateur radio networks.

   Advantages of Analog Radios:

   - Simplicity: Easier to use, especially for beginners.
   - Compatibility: Works well with older radio systems and equipment.
   - Cost: Generally more affordable.

   Drawbacks of Analog Radios:

   - Quality degrades with distance or interference.
   - Limited features compared to digital radios.

2. **Digital Radios**:
   Digital radios convert voice signals into data, which is then transmitted as a series of ones and zeroes. Digital signals are more resistant to interference and can maintain clear voice quality over longer distances. These radios often include features like encryption, data transfer, and text messaging, which aren't possible with analog radios.

   Advantages of Digital Radios:

   - Clearer audio quality over longer distances.
   - More features such as encryption and text messaging.
   - Greater security due to digital encoding.

   Drawbacks of Digital Radios:

   - Typically more expensive.
   - Complexity: Can be harder to use for beginners.
   - Less compatibility with older, analog systems.

For most amateur radio users and preppers, analog radios, like Baofeng's UV-5R, are more than adequate for reliable communication, making them a popular choice due to their simplicity and affordability.

Key Radio Terms You Should Know

To effectively use your Baofeng radio, you'll need to become familiar with some key radio terms. These terms will help you understand how your radio operates and how to communicate effectively.

- **Frequency**: The specific range of radio waves your radio transmits and receives on. Frequencies are measured in hertz (Hz), but for radios, we usually talk about megahertz (MHz). Baofeng radios can operate on both the VHF (Very High Frequency) and UHF (Ultra High Frequency) bands. VHF is better for long-distance communication in open areas, while UHF is better for shorter distances and urban environments.
- **Channel**: A specific frequency (or set of frequencies) that your radio is set to communicate on. Think of channels as predefined frequencies that make it easier for users to tune into the same "conversation."
- **Modulation**: The process of encoding your voice onto a radio wave. There are different types of modulation, such as AM (Amplitude Modulation) and FM (Frequency Modulation), which affect how the signal is transmitted. Most Baofeng radios use FM, which is better for clear voice transmission over short distances.
- **Tone Squelch (CTCSS/DCS)**: These systems allow users to filter out unwanted transmissions on the same frequency. This is useful for ensuring that you only hear signals from other radios in your group, rather than picking up every transmission on that frequency.
- **Repeater**: A device that receives a radio signal and retransmits it at a higher power, effectively extending the range of your communication. Repeaters are commonly used in amateur radio networks to cover larger areas.
- **Transmitting Power (Wattage)**: The strength of your radio's signal is measured in watts. Baofeng radios typically offer adjustable power levels, such as 1W (low) and 5W (high). Higher wattage provides more range but drains the battery faster.

Why Choose Baofeng Radios?

Now that you have a basic understanding of how radios work, you may wonder why Baofeng radios, particularly the UV-5R, are so popular. Here's why these radios have become a favorite among radio enthusiasts, preppers, and ham operators.

1. **Affordability**
   The most significant selling point of Baofeng radios is their price. You can buy a Baofeng UV-5R for around $30 to $50, whereas comparable radios from brands like Yaesu or Kenwood can cost upwards of $150 or more. This low price point makes Baofeng radios accessible to beginners and those who want to add multiple units to their emergency kits without breaking the bank.

2. **Features Packed Into a Small Package**
   Despite the low cost, Baofeng radios are loaded with features. The UV-5R, for instance, is a dual-band radio, meaning it can operate on both VHF (136-174 MHz) and UHF (400-520 MHz) frequencies. Other notable features include:
   - Dual-watch mode, allowing you to monitor two frequencies at once.
   - Programmable channels, which let you quickly switch between pre-set frequencies.
   - An FM radio tuner for listening to local FM stations.
   - A built-in flashlight for emergencies.

3. **Customizability**
   Baofeng radios are highly customizable. You can swap out antennas for better range, use external microphones for hands-free operation, and upgrade to higher-capacity batteries. Additionally, Baofeng radios can be programmed with specific frequencies using CHIRP software, making it easy to set them up for your needs.

4. **Compatibility**
   Baofeng radios work well with other two-way radios, whether they're Baofeng models or radios from other manufacturers. This is particularly useful if you're operating in a group or community where different people may have different radios.

5. **Dual-Band Operation**
   The ability to use both VHF and UHF frequencies gives Baofeng radios added versatility. VHF is ideal for long-distance communication in rural or open areas, while UHF works better in urban settings or when communicating through obstacles like buildings or trees. This dual-band flexibility makes the UV-5R a versatile choice for a wide range of environments.

Pros and Cons Compared to Other Radios on the Market

While Baofeng radios offer many advantages, they aren't perfect. Here's a comparison of their strengths and weaknesses:

Pros:

- **Affordability**: No other radio offers such a low price for the feature set.
- **Ease of Use**: Perfect for beginners or those new to radio communication.

- **Durable**: Surprisingly rugged for the cost, making them great for outdoor use.
- **Programmable**: The ability to program frequencies and settings through software is a huge plus.

Cons:

- **Audio Quality**: While Baofeng radios work well, their sound quality can be inferior to more expensive brands, especially in noisy environments.
- **Build Quality**: Though durable, Baofeng radios don't have the same build quality as higher-end models. Buttons and dials may feel less sturdy.
- **Limited Range**: The range of Baofeng radios, while adequate, may not match that of higher-wattage radios from other brands.

Conclusion

Baofeng radios, particularly the UV-5R, are a fantastic entry point for anyone looking to get into two-way radio communication. Whether you're preparing for an emergency, getting started in ham radio, or simply need a reliable communication tool for outdoor adventures, Baofeng radios offer an affordable, feature-packed solution.

In the next chapter, we'll dive deeper into how to program and set up your Baofeng radio to maximize its performance and ensure you're ready for any situation that comes your way.

Troubleshooting Malfunctions and Common Problems with Baofeng Radios

Problem	Possible Causes	Tips and Tricks for Resolution
Radio Won't Turn On	- Dead battery - Battery not seated correctly	- Charge the battery fully. - Ensure the battery is locked in place.
Poor Reception	- Incorrect frequency - Obstruction (buildings, terrain) - Poor antenna connection	- Double-check the frequency settings. - Move to a higher location or clear area. - Ensure the antenna is securely attached.
No Transmissions	- Transmitting on the wrong frequency - Low battery - Signal interference	- Verify you are on the correct frequency. - Replace or charge the battery. - Move to a different location to avoid interference.
Static or Noise	- Poor signal - Nearby electronic devices - Interference from other users	- Change your location to improve signal quality. - Turn off nearby electronic devices. - Try adjusting the squelch setting.
Battery Draining Quickly	- Backlight too bright - Continuous scanning - High power settings	- Reduce backlight brightness in settings. - Disable scanning unless necessary. - Use lower power settings when possible.
Unable to Program Frequencies	- Software issues - Incorrect programming method	- Ensure you are using the correct software (like CHIRP). - Follow programming instructions carefully.
Keypad Not Responding	- Lock feature enabled - Battery issues	- Check if the keypad lock is on; unlock if necessary. - Replace or recharge the battery if low.
Echo or Feedback When Transmitting	- Too close to another radio - Poor microphone positioning	- Maintain a distance from other radios during use. - Adjust microphone positioning for better audio clarity.
Receiving Only One Side of a Conversation	- Poor signal strength - Incorrect frequency	- Ensure both parties are on the same frequency. - Check signal strength and try adjusting the antenna.
Error Messages on Display	- Firmware issues - Software conflicts	- Perform a factory reset if necessary. - Ensure the radio firmware is up to date.
Difficulty Hearing Others	- Volume too low - Audio settings incorrect	- Adjust the volume knob to an appropriate level. - Check audio settings in the menu.
Programming Errors	- Conflicting channel settings - Incorrect CTCSS/DCS settings	- Verify settings and ensure no conflicts between channels. - Double-check CTCSS/DCS codes if used.

Tips for Beginners:

1. **Familiarize with User Manual**: Always have the user manual handy; it contains essential information on troubleshooting and operation.
2. **Practice Basic Functions**: Before heading into emergency scenarios, practice turning the radio on/off, changing channels, and making a simple transmission in a safe environment.
3. **Join Online Communities**: Participate in forums or local ham radio clubs to gain insights and tips from experienced users.
4. **Keep Spare Batteries**: Having extra, fully charged batteries on hand can prevent issues during prolonged use.
5. **Regular Maintenance**: Periodically check the radio for any wear and tear, especially the antenna and battery contacts.

These troubleshooting steps and tips should help beginners feel more comfortable using their Baofeng radios and address common issues that may arise.

Chapter 4: American Radio Regulations and Legal Use

When using Baofeng radios or any other two-way communication devices in the United States, it's essential to understand the legal landscape. The Federal Communications Commission (FCC) oversees radio communications and enforces regulations to ensure the airwaves remain orderly and safe for everyone. This chapter will explore the key regulations, licensing requirements, and the potential consequences of operating outside the law. Understanding these rules will help you use your Baofeng radio responsibly and legally.

FCC Regulations: An Overview

The **Federal Communications Commission (FCC)** is an independent U.S. government agency that regulates interstate and international communications by radio, television, wire, satellite, and cable. The FCC's role in radio communication is to allocate the radio spectrum to different types of users and services, such as public safety, commercial broadcasting, and amateur radio.

For amateur radio operators and preppers using Baofeng radios, FCC regulations ensure that users don't interfere with essential communications like those used by emergency services or government agencies. The FCC also regulates the power levels and frequencies on which different types of radios can operate.

Baofeng radios, such as the UV-5R, are highly versatile and capable of operating on various frequencies, including those reserved for ham (amateur) radio, GMRS (General Mobile Radio Service), and other services. However, it's important to know which frequencies are open for unlicensed use and which require a license to operate legally.

Rules Governing Ham (Amateur) Radio Operations in the U.S.

Ham radio, also known as amateur radio, is a popular hobby and essential communication tool during emergencies. Ham radio operators use designated frequencies to communicate, experiment with radio technology, and even support emergency services when needed. However, operating a ham radio without proper authorization is illegal in the U.S.

Here are the key rules that govern ham radio operations:

1. **Licensing**: To transmit on ham radio frequencies, you must hold an FCC-issued amateur radio license. There are three levels of licenses in the U.S.: Technician, General, and Extra. Each level grants access to more frequencies and power, but the entry-level Technician license is sufficient for most Baofeng radio users.
2. **Frequency Allocation**: Ham radio operators have access to specific portions of the VHF and UHF bands, as well as other high-frequency (HF) bands. Different license levels allow you to operate on different frequencies. For example, a Technician license gives access to popular bands such as the 2-meter (144-148 MHz) and 70-centimeter (420-450 MHz) bands, both of which Baofeng radios can access.
3. **Power Limits**: Ham radio operators must follow power restrictions to avoid interference. For instance, on the 2-meter and 70-centimeter bands, amateur radio users can transmit at higher power levels than on unlicensed bands, but they must still adhere to FCC limits.
4. **Call Signs and Identification**: Licensed amateur radio operators must identify themselves using their assigned call signs at the beginning and end of each communication, as well as at least once every ten minutes during ongoing transmissions.

Licensing Requirements

Operating on amateur radio frequencies without a license is illegal and can result in fines or other penalties. Here's a step-by-step guide to obtaining a license to legally operate your Baofeng radio on ham radio frequencies:

1. **Technician License**
   - **Description**: This is the entry-level license and allows you to operate on the VHF and UHF bands, including the popular 2-meter and 70-centimeter bands. It's the most relevant for Baofeng radio users, as it covers the frequencies these radios typically use.
   - **Exam**: You must pass a 35-question multiple-choice exam covering basic radio theory, regulations, and safety practices. The exam fee is around $15.
   - **Privileges**: You'll have access to all frequencies above 30 MHz, as well as some limited privileges on the HF bands.
2. **General License**
   - **Description**: The General license opens up more HF frequencies, allowing you to communicate over long distances, which is particularly useful for international communications.
   - **Exam**: You must pass a second 35-question exam, in addition to holding a Technician license.
   - **Privileges**: This license grants broader access to both HF and VHF/UHF bands, giving you the ability to communicate worldwide.

3. **Extra License**
   - ○ **Description**: This is the highest level of amateur radio license and grants full access to all amateur bands.
   - ○ **Exam**: You must pass a 50-question exam on advanced radio theory and operating practices, after first obtaining the General license.
   - ○ **Privileges**: As an Extra license holder, you'll have the maximum frequency privileges and priority access on many popular amateur bands.

What You Can Legally Do Without a License

For users who don't want to go through the licensing process, there are still legal ways to use your Baofeng radio. While you won't be able to access ham radio frequencies, you can operate on certain unlicensed bands designed for personal or family communication.

1. **GMRS (General Mobile Radio Service)**
   - ○ **License Required**: Yes, but no exam. The FCC requires a license for GMRS use, but the good news is that it's relatively easy to obtain—you simply pay a fee (around $35) for a 10-year license.
   - ○ **Frequencies**: GMRS operates on frequencies in the UHF band (462-467 MHz), which many Baofeng radios can access. This is a great option for family communication and emergency use.
   - ○ **Power Limits**: GMRS allows higher power than other unlicensed services, typically up to 50 watts.
2. **FRS (Family Radio Service)**
   - ○ **License Required**: No. FRS is designed for short-range, personal communication and does not require a license.
   - ○ **Frequencies**: FRS operates on the same frequencies as GMRS (462-467 MHz), but with stricter power limits (up to 2 watts).
   - ○ **Power Limits**: FRS radios are limited to lower power output, which means shorter communication ranges, typically under 1 mile.
3. **MURS (Multi-Use Radio Service)**
   - ○ **License Required**: No.
   - ○ **Frequencies**: MURS operates on VHF frequencies (151-154 MHz), offering better range in open areas compared to FRS.
   - ○ **Power Limits**: Limited to 2 watts, but you can use external antennas for better range.

These unlicensed bands provide a legal way for users to communicate without the need for an amateur radio license, making them a great option for family communication or small groups during outdoor activities.

Understanding Penalties for Misuse

The FCC takes radio regulations seriously, and violations can result in significant penalties. Operating on restricted frequencies without a license or exceeding power limits can lead to enforcement actions, including fines, equipment confiscation, and even criminal charges in extreme cases. Here are some key consequences to be aware of:

1. **Fines**: The FCC can impose fines for operating without a license or violating other radio regulations. Fines can range from a few hundred dollars to tens of thousands of dollars, depending on the severity of the violation.
2. **Equipment Confiscation**: If you're caught using a Baofeng radio illegally, the FCC may confiscate your equipment as part of their enforcement actions.
3. **Criminal Charges**: In rare cases, severe violations—especially those that interfere with public safety communications—can result in criminal charges, including imprisonment.

Examples of Misuse and Enforcement in the U.S.

There have been several instances where individuals or groups faced consequences for misusing radios in the U.S. Some notable cases include:

1. **Unauthorized Use of Emergency Frequencies**: In 2020, a ham radio operator in California was fined $25,000 by the FCC for repeatedly interfering with emergency communications during wildfires. He was caught transmitting on frequencies used by first responders, which disrupted crucial disaster relief efforts.
2. **Illegal GMRS Use**: A Texas resident was fined $7,000 in 2018 for operating on GMRS frequencies without a license. Despite warnings, the individual continued to use the frequency without authorization, leading to enforcement actions.
3. **Operating Without a License**: In New York, an unlicensed radio operator was fined $10,000 for transmitting on amateur frequencies reserved for licensed operators. The individual's transmissions interfered with local ham radio networks, leading to complaints and an FCC investigation.

These examples highlight the importance of following FCC regulations. While Baofeng radios offer incredible flexibility, it's crucial to stay within the legal bounds to avoid hefty fines or other penalties.

Conclusion

Understanding and following FCC regulations is vital for anyone using Baofeng radios in the United States. Whether you're a licensed ham radio operator or using unlicensed services like FRS and MURS, adhering to the rules ensures that you stay compliant while communicating safely and effectively. In the next chapter, we'll dive into how to program your Baofeng radio for legal use and explore some best practices for setting up channels and frequencies.

Chapter 5: Baofeng UV-5R Overview and Features

The Baofeng UV-5R is one of the most popular and affordable handheld radios in the amateur radio community. Its versatility, robust feature set, and ease of use make it a favorite among ham radio operators, preppers, and outdoor enthusiasts alike. In this chapter, we'll explore the anatomy of the Baofeng UV-5R, the power options available, and its key features, including dual-band operation, dual-watch mode, and FM radio functionality.

The Anatomy of a Baofeng Radio

Before diving into its functionality, it's important to understand the basic components of the Baofeng UV-5R. Knowing what each button and part does will help you operate your radio more efficiently, especially in emergency situations.

1. Key Components:

- **LCD Screen**: The small, backlit display shows essential information like the frequency you're tuned into, battery status, channel number, and other key settings. It's easy to read in low light and offers quick access to operational details.
- **Antenna**: The removable VHF/UHF antenna is used to send and receive signals. The stock antenna can be upgraded to improve range and performance.
- **Keypad**: The UV-5R's keypad features several buttons, including number keys, menu navigation buttons, and special function buttons. The keypad allows you to input frequencies manually, access the menu, and adjust settings on the go.
  - **PTT Button (Push-to-Talk)**: Located on the side, this is one of the most critical buttons. Pressing it allows you to transmit your voice over the radio.
  - **A/B Button**: This button allows you to switch between the two frequencies or channels you've selected for dual-watch mode.
  - **Monitor Button**: This allows you to check for weaker signals by reducing the squelch threshold, which can help you find transmissions that would otherwise be filtered out.
  - **VFO/MR Button**: This toggles between frequency mode (VFO) and memory mode (MR). In frequency mode, you can manually input a frequency, while in memory mode, you can select pre-programmed channels.
- **Volume Knob and Power Switch**: Located on the top, the volume knob also serves as the power switch. Turn it clockwise to power on the radio and adjust the volume.
- **SMA Antenna Connector**: The Baofeng uses an SMA (SubMiniature version A) connector for its antenna, making it easy to replace or upgrade.

- **Speaker and Microphone**: These are built into the front of the radio, enabling communication without additional accessories. However, you can also attach external headsets or microphones for hands-free use.

Powering Your Radio

One of the most important aspects of keeping your Baofeng UV-5R operational is ensuring it stays powered. The radio uses a rechargeable battery, and understanding how to maximize its life and efficiency is key to being prepared for both everyday use and emergencies.

1. Battery Types and Battery Life

The Baofeng UV-5R typically comes with a **1800mAh lithium-ion battery**, but extended batteries with capacities up to 3800mAh are also available for those who need longer operational times.

- **Standard 1800mAh Battery**: This provides about 12 to 18 hours of use under normal conditions. Actual battery life depends on how often you transmit, as transmitting uses significantly more power than receiving or listening.
- **Extended Battery Options**: For those who need more endurance (e.g., preppers or outdoor enthusiasts), you can purchase extended batteries that can last up to 24 to 36 hours on a single charge. These larger batteries make the radio bulkier but are excellent for prolonged field use.

2. Charging Tips and Best Practices

Proper charging habits can extend your battery's life and improve the overall performance of your Baofeng radio.

- **Charge Before First Use**: It's recommended to fully charge your Baofeng radio before using it for the first time. Most lithium-ion batteries come with a partial charge, but topping them off ensures you're starting at full capacity.
- **Avoid Overcharging**: Although modern chargers stop delivering current when the battery is full, it's still a good practice to avoid leaving your radio on the charger for extended periods (such as overnight). This helps maintain battery health over time.
- **Store Batteries Properly**: If you're not using your Baofeng radio for an extended period, it's best to store the battery at around 50% charge. Fully charged or fully discharged batteries degrade faster when left unused.

- **Battery Bank Option**: For outdoor adventurers or preppers, consider using a **solar charger** or **power bank** to keep your radio charged when away from traditional power sources. Many preppers carry an extra charged battery as a backup.

Key Features of the Baofeng UV-5R

The UV-5R's feature set is what makes it such a favorite among radio enthusiasts. Let's break down the core features that distinguish it from other radios in the market.

1. Dual-Band Operation (VHF and UHF)

One of the most attractive features of the Baofeng UV-5R is its ability to operate on both **VHF (Very High Frequency)** and **UHF (Ultra High Frequency)** bands.

- **VHF (136-174 MHz)**: VHF is great for outdoor use, especially in rural areas, as the signal can travel longer distances and penetrate obstacles like trees and buildings more effectively.
- **UHF (400-520 MHz)**: UHF is better suited for urban environments where obstacles like walls and metal structures are more common. UHF signals can bounce off obstacles, providing better performance in cities.

This dual-band capability allows users to communicate over a wide range of frequencies, making the UV-5R incredibly versatile for both ham radio operators and emergency communications.

2. Dual-Watch Mode

The UV-5R also includes a **dual-watch feature**, which allows you to monitor two frequencies or channels at the same time. This can be extremely useful in emergency situations where you need to listen to a primary emergency channel while also keeping an ear on another frequency.

- **A/B Button**: You can toggle between the two channels you're monitoring by pressing the A/B button. This allows you to switch which frequency you want to actively transmit on.

3. FM Radio Function

In addition to its two-way radio capabilities, the Baofeng UV-5R has a built-in **FM radio receiver**. This allows you to listen to regular FM broadcasts (88-108 MHz) in between transmissions.

This feature can be useful for entertainment or staying updated on news broadcasts during emergencies when regular communication channels may be down. If you're in a crisis situation, tuning in to local news stations can provide critical updates on weather conditions, natural disasters, or government instructions.

4. LED Flashlight

The Baofeng UV-5R also includes a small **LED flashlight** built into the top of the radio. While it's not the brightest flashlight, it's a handy feature in low-light situations or emergencies when you need to quickly illuminate your surroundings.

5. VOX (Voice-Activated Transmission)

Another convenient feature is the **VOX (Voice Operated Exchange)** function, which allows for hands-free operation. With VOX enabled, the radio will automatically start transmitting as soon as it detects your voice, without the need to press the PTT button. This can be especially useful in situations where both of your hands are occupied, such as when hiking, driving, or during emergencies.

6. Programmable Channels

The UV-5R comes with **128 programmable channels**, giving you the flexibility to store frequently used frequencies for quick access. Channels can be programmed directly through the keypad or using a programming cable and software (such as CHIRP), which makes organizing your frequencies much easier.

Conclusion

The Baofeng UV-5R is packed with useful features that make it a standout option for radio enthusiasts, preppers, and emergency communicators alike. Its dual-band capability, dual-watch feature, and affordability make it a versatile tool in both everyday and emergency situations. In the next chapter, we'll go through step-by-step instructions on how to program your Baofeng UV-5R, along with tips for optimizing your radio for different scenarios.

Summary of Baofeng Radio Models: Features, Usage, Battery Life, and Reviews

Model	Features	Usage	Battery Life	Reviews
Baofeng UV-5R	- Dual-band (VHF/UHF) - 128 programmable channels - Dual watch - FM radio - Emergency alarm	General use, ham radio, preppers	12-24 hours (with normal use)	Highly popular; praised for affordability and versatility. Some users note a steep learning curve.
Baofeng BF-F8HP	- High power (8W) - Improved battery (2000mAh) - Tri-color display - Dual band (VHF/UHF)	Outdoor activities, emergency prep	24-36 hours	Well-received; noted for robust build and battery life. Slightly more expensive than UV-5R but worth it for the features.
Baofeng UV-82	- Dual push-to-talk buttons - 5W power - Enhanced audio quality - Dual band (VHF/UHF)	Commercial use, security, outdoor events	12-20 hours	Generally positive; appreciated for audio clarity and dual PTT for easy operation. Some find it bulkier than UV-5R.
Baofeng UV-5RA	- Similar to UV-5R - 5W power - VHF/UHF - Emergency alarm	General ham use, beginner-friendly	12-24 hours	Good reviews for newcomers; affordable option but with standard features.
Baofeng UV-5R+	- Enhanced UV-5R features - Dual watch - FM radio - 5W power	General use, emergency services	12-24 hours	Users like the upgrades from the standard UV-5R; still affordable.
Baofeng GT-3TP	- High power (8W) - Dual band (VHF/UHF) - IP67 waterproof	Outdoor adventures, emergencies	24-36 hours	Excellent durability; users report great performance in tough conditions. Higher

Model	Features	Usage	Battery Life	Reviews
	rating - 2000mAh battery			price but justified for features.
Baofeng UV-5X3	- Triple-band (VHF/UHF/220 MHz) - 6W power - Large color display - 2000mAh battery	Ham radio enthusiasts	24-36 hours	Niche model praised for versatility in band coverage; relatively new with limited reviews.
Baofeng DM-1701	- DMR (Digital Mobile Radio) support - Dual band (VHF/UHF) - 5W power	Digital communication, ham use	15-25 hours	Positive feedback for digital features; some find the setup process complicated.
Baofeng UV-50X3	- Triple-band (VHF/UHF/220 MHz) - Dual watch - 50W power	Advanced amateur radio, emergencies	20-30 hours	High-performance model; users appreciate the power but note it may require advanced knowledge.
Baofeng UV-9R	- IP67 waterproof - Floatable design - 8W power - Dual band (VHF/UHF)	Marine use, extreme conditions	20-30 hours	Good reviews for ruggedness; users like the waterproof feature but some mention complexity in operation.

Notes:

- **Battery Life**: Values are approximations based on typical use; actual battery life can vary with usage conditions, power settings, and features in use.
- **Usage**: Indicates common scenarios for which each model is well-suited, but all models can generally be used for multiple applications.
- **Reviews**: Summaries are based on user feedback across various platforms, highlighting general sentiments about each model.

These models offer a range of functionalities to cater to different users, from beginners to seasoned ham radio operators, ensuring there's a suitable option for anyone interested in two-way communication.

Chapter 6: Programming Your Baofeng Radio

Programming your Baofeng UV-5R is essential for getting the most out of your radio, whether you're using it for ham radio, emergency preparedness, or general communication. While the radio can be used straight out of the box, programming it correctly ensures you have access to the frequencies you need in various scenarios. In this chapter, we'll cover both **manual programming** and programming with the **CHIRP software** to make your Baofeng radio ready for any situation. We'll also discuss some recommended frequencies for U.S. users.

Manual Programming

Manually programming your Baofeng radio is a useful skill, especially in situations where you don't have access to a computer or programming cable. Learning how to input frequencies and save channels on the go is crucial for effective communication in the field.

Step-by-Step Instructions for Manual Programming

Here's how to manually program a frequency into your Baofeng UV-5R:

1. **Switch to Frequency Mode (VFO)**
   - Press the **VFO/MR button** to switch from Memory mode (MR) to Frequency mode (VFO). In Frequency mode, you can input frequencies directly.
2. **Enter the Frequency**
   - Use the keypad to enter the desired frequency. For example, to enter 146.520 MHz (the national simplex calling frequency), type **1-4-6-5-2-0**.
3. **Set the Transmit (TX) Frequency**
   - If the frequency requires a different transmit frequency (for a repeater, for example), press **MENU** and use the arrow keys to find **Menu 25: SFT-D** (Offset Direction). Set it to + or - depending on the repeater's offset direction.
   - Next, go to **Menu 26: OFFSET** to set the offset. The most common offset for VHF repeaters is 0.600 MHz, so enter **0-0-6-0-0**.
4. **Set CTCSS/DCS Tones**
   - Some repeaters require a **CTCSS** or **DCS tone** to access. To set this, press **MENU** and go to **Menu 13: T-CTCS** (for transmit tone) or **Menu 11: R-CTCS** (for receive tone). Select the appropriate tone frequency, which you can usually find in your repeater's directory.
5. **Save the Frequency to a Channel**

o Once you've entered all the necessary settings, press **MENU** and go to **Menu 27: MEM-CH** to assign the frequency to a channel. Enter a channel number that's not already in use (e.g., 001) and press **MENU** to save.

6. **Name the Channel (Optional)**
   o Baofeng radios don't allow you to name channels directly from the device, but if you use CHIRP software (discussed below), you can give each channel a name for easier identification.

Deleting Channels

If you need to free up space or remove a channel you no longer use, follow these steps:

1. Press the **VFO/MR button** to switch to Memory mode (MR).
2. Press **MENU** and go to **Menu 28: DEL-CH**.
3. Use the arrow keys to select the channel you want to delete, then press **MENU** to confirm.

Using CHIRP Software

While manual programming is useful in certain situations, using software to program your Baofeng radio can save you time and allow you to manage more complex setups, like repeater lists, tones, and channel names. **CHIRP** is a free, open-source software program widely used by radio operators to program radios, including the Baofeng UV-5R.

How to Download and Install CHIRP

Follow these steps to get started with CHIRP:

1. **Download CHIRP**
   o Visit the official CHIRP website (chirp.danplanet.com) and download the latest version of the software that matches your operating system (Windows, Mac, or Linux).
2. **Install CHIRP**
   o Follow the installation instructions provided for your operating system. For Windows users, you might need to install additional drivers to recognize the USB programming cable.
3. **Connect Your Baofeng Radio**

o Use a compatible USB programming cable (usually a USB to Kenwood 2-pin cable) to connect your radio to the computer. Make sure your Baofeng radio is turned on.

4. **Import Your Radio Settings**
   o Once CHIRP is open, click **Radio > Download From Radio**. Select the correct COM port (this is automatically detected in most cases) and the Baofeng UV-5R model. CHIRP will import your current settings from the radio.

Importing Frequency Lists and Advanced Settings

One of the best features of CHIRP is the ability to import pre-made frequency lists and set advanced settings, such as power levels, squelch, and more.

1. **Import Frequency Lists**
   o You can find pre-made frequency lists online for your area, including repeaters, local prepper networks, and emergency services. Import them into CHIRP by selecting **File > Import** and choosing the appropriate .csv file.
2. **Set Channel Names**
   o Under the **Memories** tab, you can assign names to each channel, such as "Local Repeater" or "Emergency Freq." This makes it easier to identify your channels when you're out in the field.
3. **Save and Upload Settings**
   o After making any changes, select **Radio > Upload to Radio** to transfer the new settings back to your Baofeng UV-5R. This will overwrite the previous settings, so make sure everything is correct before uploading.

Recommended Frequencies

Now that you know how to program your Baofeng UV-5R, it's time to populate it with useful frequencies. Below are some of the most common and essential frequencies for U.S. users, especially for ham operators and preppers.

1. Local Repeaters

Most ham radio operators will rely on local repeaters to extend their communication range. You can find a list of repeaters in your area on websites like **RepeaterBook** or by consulting a local ham club. Here are some common repeater frequencies for reference:

- **146.520 MHz** (National simplex calling frequency, VHF)
- **446.000 MHz** (National simplex calling frequency, UHF)
- **144.390 MHz** (APRS frequency for tracking and messaging)

2. Emergency Services

While most emergency service frequencies are restricted, certain public service frequencies may still be useful to monitor in a disaster or emergency situation:

- **162.550 MHz** (NOAA Weather Radio)
- **155.160 MHz** (EMS dispatch)
- **462.675 MHz** (GMRS emergency frequency)

3. Prepper Networks

Preppers often use specific frequencies for regional and national communications. Some frequencies commonly used by preppers include:

- **3.818 MHz** (Prepper net frequency on HF)
- **7.242 MHz** (Prepper HF network)

4. FRS/GMRS/MURS

For users without a ham license, FRS and GMRS frequencies offer legal communication options without a need for advanced licensing (GMRS does require a license, but no exam). These frequencies are useful for short-range communication during events or emergencies:

- **462.5625 MHz** (FRS/GMRS shared channel 1)
- **151.820 MHz** (MURS channel 1)

Conclusion

Programming your Baofeng UV-5R is a critical step in getting the most out of your radio, especially if you're planning to use it for emergencies or to communicate with local ham radio operators. Whether you're manually inputting frequencies or using CHIRP software for advanced settings, a well-programmed radio is a powerful tool. In the next chapter, we'll dive deeper into advanced communication techniques and emergency scenarios where your Baofeng can be a lifesaver.

Chapter 7: Operating Your Baofeng Radio

Now that your Baofeng UV-5R is programmed and ready, it's time to learn how to operate it efficiently. In this chapter, we'll walk you through the basic operation of your radio, including how to power it on, adjust volume, and switch channels. We'll also cover the essentials of transmitting and receiving, proper radio etiquette for amateur operators, and how to use the scan function to find active frequencies.

Basic Operation

Turning On the Radio

1. **Power On**: To turn on your Baofeng UV-5R, twist the **volume knob** (located on the top of the radio) clockwise. As the radio powers up, you'll hear a beep and the LCD screen will light up. You'll also see the last frequency or channel used displayed on the screen.
2. **Adjusting the Volume**: Continue twisting the volume knob clockwise to increase the volume, or counterclockwise to decrease it. Test the volume by tuning into an active frequency or by using the FM radio feature to make sure you can hear clearly.

Switching Channels

Once your radio is on, you'll want to switch between the channels or frequencies you've programmed. There are two primary modes for operating the Baofeng UV-5R: **Frequency Mode (VFO)** and **Memory Mode (MR)**.

- **Memory Mode (MR)**: In this mode, you can easily switch between pre-programmed channels. Press the **VFO/MR button** to enter Memory mode, then use the **arrow keys** to scroll through your saved channels. The screen will display the channel number, and if you've named the channels using CHIRP software, it will also show the channel name.
- **Frequency Mode (VFO)**: If you need to manually input a frequency, press the **VFO/MR button** to switch to Frequency mode. Use the keypad to enter the frequency you want to monitor or transmit on. This mode is useful when you need to communicate on an unlisted frequency or test a new one.

Transmitting and Receiving

The most important function of any two-way radio is transmitting and receiving. To communicate effectively, it's essential to understand the proper techniques and radio etiquette.

How to Make a Transmission

1. **Select the Right Frequency or Channel**: Before transmitting, make sure you're on the correct frequency or channel, and that it's not already in use. You can monitor the channel for a moment to ensure it's free.
2. **Press the Push-to-Talk (PTT) Button**: On the left side of your Baofeng UV-5R, you'll find the **PTT button**. Press and hold it to begin your transmission. The LCD screen will show a red indicator, signaling that you are actively transmitting. Be sure to hold the radio a few inches away from your mouth and speak clearly.
3. **Release the PTT Button**: Once you've finished speaking, release the PTT button. The radio will switch back to receiving mode, and you'll see the indicator on the LCD turn back to green when a signal is received.

Radio Etiquette for Amateur Operators

Following proper radio etiquette is especially important when communicating over amateur frequencies. Here are a few key guidelines:

- **Listen First**: Before transmitting, always listen to the channel to ensure it's clear. Never interrupt an ongoing conversation unless it's an emergency.
- **Use Call Signs**: If you're a licensed ham operator, always identify yourself by using your **call sign** at the beginning and end of your transmission. For example: "This is **KD9XYZ**."
- **Keep Transmissions Brief**: Only transmit the necessary information to keep the channel clear for other users. Long-winded transmissions can be frustrating for others waiting to use the frequency.
- **Avoid Slang or Code Words**: Stick to clear, concise language. Avoid using jargon, CB slang, or "10-codes" unless they are universally understood by your communication group.

Transmitting on FRS/GMRS/MURS

If you're transmitting on FRS, GMRS, or MURS channels, be sure to follow the appropriate etiquette and licensing rules. FRS requires no license, but GMRS requires a license from the

FCC. Always ensure you are using the proper power settings and channel designations to stay compliant.

Scanning for Active Channels

The **scan function** on your Baofeng UV-5R is a handy feature for finding active channels in your area. Whether you're looking for local repeater traffic, emergency communications, or prepper networks, scanning helps you locate frequencies that are in use.

How to Use the Scan Function

1. **Activate Scanning**: To start scanning, press and hold the **SCAN button** (which is usually the *key* button or an assigned soft key on your radio) until the radio begins scanning through the frequencies or channels. The radio will stop momentarily on any active frequency it detects.
2. **Monitor an Active Frequency**: When the radio stops on an active frequency, it will remain on that frequency until the transmission ends. If you want to stay on that channel, simply press any button on the keypad to stop scanning.
3. **Resume Scanning**: If you want to continue scanning, press the **SCAN button** again. The radio will resume scanning until it finds the next active channel.
4. **Adjust Scan Settings**: You can adjust the scan settings (such as the squelch level) to filter out weak or noisy transmissions. Lower squelch settings allow you to pick up weaker signals, while higher settings block out noise.

Tips for Efficient Scanning

- **Focus on Key Bands**: If you're only interested in certain frequencies (like VHF or UHF bands), make sure you limit your scan range to those frequencies.
- **Program Local Repeaters**: Make sure you've programmed all local repeaters into your memory channels. This will allow your radio to quickly scan those channels for activity.
- **Emergency Scanning**: During an emergency, scanning can be a valuable tool for finding live emergency broadcasts or communicating with nearby operators.

Conclusion

Operating your Baofeng UV-5R doesn't need to be complicated. With a basic understanding of how to turn the radio on, switch channels, and transmit and receive properly, you'll be able to use your Baofeng effectively in a variety of situations. The scan function is also a powerful tool for finding active frequencies when you need to stay informed or communicate in real-time.

In the next chapter, we'll explore advanced communication techniques, such as using repeaters, setting up emergency channels, and maximizing your radio's range in difficult environments.

Chapter 8: Frequency Bands and Their Uses

Understanding the different frequency bands and their uses is essential for effectively operating your Baofeng radio, especially when it comes to optimizing range, clarity, and versatility in various settings. In this chapter, we'll explore the distinctions between **VHF** (Very High Frequency) and **UHF** (Ultra High Frequency) bands, outline their common applications, and highlight some of the key American radio frequencies, including essential emergency and weather channels.

VHF (Very High Frequency)

The VHF band, which ranges from 30 MHz to 300 MHz, is widely used for **long-distance outdoor communication** and is popular among amateur radio operators, marine operators, and preppers due to its extended range in open environments.

Common Uses of VHF

1. **Amateur Radio Repeaters**: VHF frequencies (typically around 144–148 MHz) are frequently used in amateur radio, especially with repeaters. Repeaters extend the range of VHF communication, making it possible to connect over greater distances, even with obstacles like hills or valleys.
2. **Marine Communication**: VHF is also designated for marine operations, including ship-to-shore communication and emergency calls at sea, owing to its reliable performance over open water.
3. **Outdoor and Rural Communication**: Due to its ability to travel long distances over open spaces, VHF is favored for communication in rural or outdoor areas where few obstructions interfere.

Coverage and Range of VHF

- **Open Terrain**: VHF works exceptionally well in open spaces, covering up to several miles when unimpeded by buildings or dense forests.
- **Hilly Areas**: While VHF can handle moderate terrain, its signals are generally less effective in mountainous areas without repeaters.

- **Urban or Indoor Environments**: VHF is more likely to encounter interference from buildings and walls, which can reduce its effectiveness in cities or when used indoors.

UHF (Ultra High Frequency)

Spanning from 300 MHz to 3 GHz, UHF frequencies are ideal for **shorter-range communication** in urban environments, indoors, and even around dense forests. This makes UHF popular for **GMRS**, **FRS**, and **indoor operations**, where clear communication over shorter distances is essential.

Typical Applications for UHF

1. **Urban and Indoor Communication**: UHF's shorter wavelengths allow it to penetrate walls and buildings more effectively than VHF, making it preferable for urban settings and indoor communication.
2. **GMRS and FRS Channels**: General Mobile Radio Service (GMRS) and Family Radio Service (FRS) primarily operate on UHF frequencies, making them ideal for close-range communication. GMRS requires a license, while FRS can be used without one.
3. **Personal and Prepper Use**: UHF's strong performance in cluttered environments makes it useful for preppers who need reliable communication in areas with buildings, forests, or other obstacles.

Coverage and Range of UHF

- **Urban Settings**: UHF excels in urban areas, as its signals penetrate buildings and walls with minimal interference.
- **Dense Forests and Suburban Areas**: UHF can handle environments with a moderate amount of physical obstructions, although thick forests can limit range somewhat.
- **Open Terrain**: While UHF's shorter wavelengths limit its range in open spaces, it still provides clear audio quality within a limited radius.

Important American Radio Frequencies

To make the most out of your Baofeng, it's essential to know some key frequencies used across different services in the U.S. These include FRS, GMRS, MURS channels, as well as critical emergency and weather frequencies.

1. General Mobile Radio Service (GMRS)

- **462.550 MHz to 462.725 MHz**: GMRS channels are used for short-distance family communication, particularly in outdoor or suburban areas. They provide better range than FRS channels but require an FCC license.

2. Family Radio Service (FRS)

- **462.5625 MHz to 462.725 MHz**: FRS shares frequencies with GMRS but has lower power limits, making it suitable for personal communication over shorter distances. It's license-free and commonly used by families and recreational groups.

3. Multi-Use Radio Service (MURS)

- **151.820 MHz to 154.600 MHz**: MURS channels are popular among preppers and small businesses. These frequencies don't require a license and offer a decent range without interference from repeaters or public traffic.

Emergency and Prepper Frequencies

In emergency situations or for prepper communications, having dedicated frequencies can make a significant difference. Below are some commonly used frequencies for these purposes.

- **3.818 MHz (HF)**: Used by preppers for long-distance communication, particularly at night, when HF frequencies propagate better.
- **7.242 MHz (HF)**: Another prepper frequency, used mainly during the day for regional communication over several hundred miles.
- **146.520 MHz (VHF)**: The national calling frequency for ham radio simplex operation. This is monitored by ham operators across the country for local communication.
- **446.000 MHz (UHF)**: The UHF national calling frequency. This frequency is monitored for simplex communication among operators in urban and suburban areas.

NOAA Weather Radio Frequencies

For real-time weather alerts and warnings, the National Oceanic and Atmospheric Administration (NOAA) broadcasts on a set of dedicated frequencies. These frequencies cover most of the U.S. and provide 24/7 weather updates.

- **162.400 MHz to 162.550 MHz**: These seven frequencies are reserved for NOAA weather broadcasts. Most regions have at least one NOAA station that covers local weather, storm updates, and emergency alerts. Monitoring these channels ensures you stay informed about changing weather conditions, especially in disaster-prone areas.

Essential Emergency Frequencies for Baofeng Radio

Frequency	Service	Purpose	Notes
146.520 MHz	Amateur (2m band)	National Calling Frequency	Simplex calling frequency for emergencies
146.420 MHz	Amateur (2m band)	Prepper Network / Backup	Often used in local prepper groups
462.675 MHz	GMRS (Channel 20)	General Emergency	Requires GMRS license; used in emergencies
462.550 MHz	GMRS (Channel 15)	Emergency Backup	Common GMRS emergency frequency
151.940 MHz	MURS (Multi-Use Radio Service)	General Communication	License-free; used by preppers
154.570 MHz	MURS (Multi-Use Radio Service)	Emergency Communication	License-free; reliable for short-range
154.600 MHz	MURS (Multi-Use Radio Service)	Emergency Backup	Good for short-range, no license required
121.500 MHz	Aircraft Emergency	Distress Signal	Use only in extreme emergencies
162.400 - 162.550 MHz	NOAA Weather Radio	Weather Alerts	NOAA broadcasts for severe weather info
446.000 MHz	Amateur (70cm band)	National Simplex Frequency	Common emergency and calling frequency
156.800 MHz	Marine VHF (Channel 16)	Distress and Calling (Coast Guard)	For water-related emergencies only
467.7125 MHz	FRS Channel 22	Emergency Communication (Prepper Groups)	License-free for FRS users

Note:

- **Amateur (Ham) Frequencies** require an FCC license (Technician or higher).
- **GMRS Frequencies** require a GMRS license, while **FRS** and **MURS** channels are license-free for personal use.
- **Marine and Aircraft frequencies** are restricted for their designated uses.

Conclusion

Choosing the correct frequency band and knowing the relevant American radio frequencies for communication needs—from general use to emergency alerts—is crucial for operating your Baofeng effectively. Understanding the strengths of **VHF** for outdoor and long-range use and **UHF** for urban and indoor settings will help you maximize the radio's capabilities. By familiarizing yourself with these essential frequencies, you'll be prepared for daily communication, recreational activities, and even emergency situations.

In the next chapter, we'll delve into advanced tips and techniques to boost your radio's performance, helping you extend range, improve audio clarity, and get the most out of your Baofeng in various conditions.

Chapter 9: Accessories and Upgrades

While Baofeng radios are powerful tools on their own, adding the right accessories can significantly enhance their performance, usability, and longevity. In this chapter, we'll explore some key accessories, including antenna upgrades for improved range, external microphones and headsets for hands-free use, and other practical additions like battery packs, charging options, and protective cases.

Antennas

One of the simplest and most impactful upgrades for a Baofeng radio is replacing the stock antenna with a higher-performance alternative. Choosing the right antenna can improve transmission range, signal clarity, and overall performance, particularly in challenging environments.

Upgrading the Stock Antenna

Most Baofeng radios come with a basic "rubber duck" antenna, which is functional but limited. By switching to a higher-gain antenna, you can significantly extend your reach. Common upgrades include **whip antennas**, which are longer and more flexible than the stock option and often boost range by several miles, especially in open or rural areas.

Types of Antennas and Their Uses

- **Whip Antennas**: These longer, flexible antennas are ideal for extended-range communication. Their design allows them to capture and transmit more signal, making them excellent for outdoor, rural, or prepper applications where you may need to reach contacts over greater distances.
- **Magnetic Mount Antennas**: Popular for mobile setups, magnetic mount antennas can be placed on metal surfaces (like a car roof), allowing you to use your Baofeng as a mobile station. This setup provides added height for your antenna, increasing your range while on the road or in urban environments.
- **Yagi Antennas**: Known for their directional power, Yagi antennas can boost performance in one specific direction. These are particularly useful if you need to communicate with a specific station or repeater at a distance, making them ideal for base station setups or fixed locations.

Emergency Antenna Options for Baofeng Radios: Design, Effectiveness, and Applications

Antenna Type	Description	Frequency Range	Advantages	Notes
Stock Antenna (Rubber Duck)	The standard antenna that comes with the Baofeng radio.	VHF: 136-174 MHz UHF: 400-520 MHz	Compact and easy to carry; decent for local communication.	Basic performance, limited range.
Extended Rubber Duck	A longer rubber duck antenna that improves range.	VHF/UHF	Increased gain; better performance than stock.	Still portable; good for emergencies.
VHF Ground Plane Antenna	Simple vertical antenna designed for VHF frequencies.	VHF: 144-148 MHz	Effective for local communications; can be home-built.	Requires a ground plane for optimal performance.
UHF Ground Plane Antenna	Similar design to the VHF version, optimized for UHF.	UHF: 430-450 MHz	Improved performance in urban environments.	Can also be home-built with basic materials.
J-Pole Antenna	A simple, effective design for VHF/UHF communication.	VHF: 144-148 MHz UHF: 440-450 MHz	High gain; good for long-range communication.	Can be built from copper wire; vertical installation needed.
Dipole Antenna	A basic antenna design that can be cut to specific frequencies.	VHF/UHF	Versatile and effective; can be made from wire.	Requires space and proper tuning.
Magnetic Mount Antenna	An antenna that can be mounted on metal surfaces.	VHF/UHF	Easy to set up; good for portable operations in vehicles.	Offers flexibility and improved reception.
Yagi Antenna	A directional antenna for targeted communication.	VHF/UHF	High gain; excellent for long-range communication.	Requires alignment; less portable.
Collinear Antenna	A vertical antenna designed for enhanced range.	VHF/UHF	Good gain; compact design; suitable for mobile use.	Provides better performance than simple verticals.
Portable Tape Measure Antenna	An inexpensive, easily constructed antenna for VHF/UHF.	VHF: 144-148 MHz	Lightweight and foldable; great for emergencies.	Requires some tools for construction.
Log-Periodic Antenna	A broadband directional antenna that covers a range of frequencies.	VHF/UHF	Effective for various frequencies; good for emergency versatility.	Can be larger and requires setup.

Notes:

- **Portability**: Consider how easy it is to transport the antenna, especially in emergency situations.
- **Construction**: Some antennas can be easily built using common materials, which is useful when resources are limited.
- **Tuning**: Many antennas will need to be tuned for optimal performance on specific frequencies.
- **Use Cases**: Each antenna type serves different scenarios, from local communication to long-range contacts. Choose based on your specific needs and environment.

External Microphones and Headsets

Enhancing usability, external microphones and headsets offer hands-free operation and greater flexibility, especially in situations where quick communication or stealth is essential.

Benefits of External Microphones and Headsets

- **Shoulder Microphones**: Often used by security professionals and first responders, shoulder microphones clip onto clothing, making it easy to speak into the microphone without needing to hold the radio. This accessory frees up your hands and allows for more comfortable, seamless operation in dynamic environments.
- **Headsets**: Available in a variety of styles, from lightweight earpieces to noise-canceling headsets, these accessories are perfect for noisy or crowded situations where clear communication is critical. Some headsets come with a push-to-talk (PTT) button on the cord, making transmission even more convenient.
- **Privacy and Discretion**: Using a headset provides privacy in crowded areas or sensitive situations, as only the user can hear incoming transmissions. This feature can be particularly valuable for preppers or those using radios for personal security.

Other Accessories

Beyond antennas and microphones, several other accessories can enhance the functionality and ease of use of your Baofeng radio.

Battery Packs

- **Extended Battery Packs**: A larger battery pack increases your radio's operating time, reducing the need for frequent recharges. Extended batteries can be particularly useful for outdoor enthusiasts, preppers, or users in emergency situations where access to charging options is limited.
- **Battery Eliminators**: Designed for mobile setups, battery eliminators connect your radio directly to a car's 12V power outlet, ensuring unlimited power for long drives or extended operations. This accessory is perfect for vehicle-based communication or when you need a stable power source.
- **Spare Batteries**: Keeping spare batteries on hand ensures that your Baofeng is always ready to go. They're especially useful in field conditions where charging isn't possible, giving you a quick solution to keep your radio operational.

Charging Stations

- **Desktop Charging Docks**: A dedicated charging station or desktop dock provides a convenient place to keep your radio charged and ready. Many models are designed to charge both the radio and a spare battery simultaneously, which is ideal for users who need their radio continuously powered.
- **Solar Chargers**: For those who operate in remote areas, a solar charger can be a game-changer. These chargers allow you to power your radio's batteries using sunlight, providing a sustainable solution for long-term off-grid use.

Protective Cases and Mounts

- **Cases and Holsters**: A rugged case or holster can protect your Baofeng from drops, dirt, and moisture, extending its lifespan. Many cases also feature clips or attachments, making it easier to carry your radio on a belt or backpack for quick access.
- **Vehicle Mounts**: For users who operate their radio while driving, a vehicle mount secures the Baofeng in place, ensuring it's accessible but not distracting. Combined with a magnetic mount antenna and battery eliminator, this setup is perfect for mobile communication setups.

Conclusion

Investing in the right accessories can make a tremendous difference in how well your Baofeng radio serves you. From enhanced range with a high-performance antenna to increased usability with a headset or shoulder mic, each upgrade contributes to a more versatile and efficient communication setup. By choosing the accessories that best fit your usage needs, you can customize your Baofeng to perform at its peak, whether for daily use, outdoor adventures, or emergency situations.

In the next chapter, we'll explore some practical scenarios where these accessories can be particularly helpful and discuss how to set up and maintain your Baofeng for different types of operations.

Chapter 10: Radio for Preppers and Survivalists

In uncertain times, effective communication is essential. For preppers and survivalists, a reliable radio like the Baofeng UV-5R can be a game-changer when other communication methods are unavailable. This chapter delves into why radios are indispensable for preparedness, the key elements of setting up a strategic communication plan, and how to adapt radio use to various disaster scenarios.

Why Radios Are Essential for Preppers

When traditional infrastructure—such as phone networks, the internet, or even power grids—goes down, a radio becomes a lifeline. Radios provide a means of connecting with family, friends, or community members, ensuring vital information and resources are shared.

Critical Communication in Grid-Down Situations

During widespread power outages or disruptions, cell towers and internet service providers are among the first systems to fail. Radios, especially handheld models, do not rely on these systems, enabling **direct, point-to-point communication** over both short and moderate distances. This feature allows preppers to maintain contact with local networks or family members without depending on a third-party infrastructure.

Linking Up with Prepper and Survivalist Networks

Another crucial use of a Baofeng radio is the ability to connect with **like-minded communities and local survivalist groups**. Many prepper groups organize regular "net checks" over specific frequencies, which serve as practice for emergency scenarios and a means to check in with each other. In a disaster situation, these networks can provide information, support, and resources to help you stay informed and prepared.

Setting Up a Communications Plan

A well-thought-out communication plan ensures that you and your group know how to stay in touch if conventional methods fail. This plan includes selecting predetermined frequencies, scheduling check-ins, and establishing backup options.

Pre-Arranged Frequencies for Group Use

- **Family and Group Channels**: Designate specific channels for family or close group use. By pre-assigning channels, each member of your household or team knows where to reach others without needing to coordinate in the moment.
- **General and Emergency Frequencies**: Along with private channels, plan to monitor open emergency frequencies, where critical updates from authorities or local networks may be broadcast. It's wise to tune in to these channels periodically for new information.

Dealing with Phone and Internet Blackouts

A grid-down situation means you need alternatives. Establish **a protocol for radio check-ins** (e.g., hourly or at specific times of day) so group members know when to listen. This method conserves battery life, which is especially valuable when recharging options may be limited.

- **Out-of-Area Contacts**: Designate a friend or family member outside the affected region who can act as a relay. Group members can try reaching this person to provide updates or request assistance if local connections are compromised.

Radio Operation in Disaster Scenarios

Whether due to natural events, technology failures, or social disruptions, a variety of scenarios can impact communication. Here's how to leverage your Baofeng radio in specific disaster contexts.

Natural Disasters

From hurricanes to earthquakes, natural disasters often damage infrastructure, including communication networks. In such cases:

- **Pre-set Key Frequencies**: Program your Baofeng with frequencies of local emergency responders and weather services (like NOAA) to access real-time updates and warnings.
- **Family Contact Protocol**: If separated, family members should know when to check in or monitor certain channels for updates. This protocol minimizes confusion and helps ensure all parties are aware of each other's status.

Electromagnetic Pulse (EMP)

An EMP can disrupt electronics, potentially damaging unshielded devices. Radios may still work if protected by storing them in a **Faraday cage** or similar shielding.

- **Alternative Power Options**: In case of an EMP, have solar chargers or manually powered options on hand to keep your devices operational. Radios stored properly can still work and enable communication, even when most modern devices have been compromised.

Civil Unrest and Security Concerns

In times of social instability, it's important to stay informed and out of harm's way. Radios enable you to avoid high-risk areas, stay updated on road closures, and connect with trusted contacts.

- **Local Monitoring**: Tune in to local frequencies to stay updated on nearby incidents or checkpoints. Local frequencies give you insights that may not be covered in broader news sources.
- **Silent Operation and Headsets**: Using a headset helps keep communication discreet. When operating a radio in sensitive scenarios, the ability to listen quietly without broadcasting your location is a significant advantage.

Using the Baofeng in the Absence of Conventional Systems

When traditional methods of communication—like cell service and internet—are down, your Baofeng radio becomes an essential tool for adapting and staying in contact.

Designated Emergency Channels and Check-Ins

Pre-program key emergency channels for nearby medical, fire, and rescue services so that you can quickly tune in as needed. Schedule routine check-ins with your group on predetermined channels, conserving battery by powering on the radio only for designated intervals.

Battery Management and Backup Power

Efficient power usage is vital in prolonged scenarios without reliable electricity. Make use of **backup battery packs** and, if possible, invest in a **solar charger** or battery eliminator. Rotate batteries to keep them charged, and consider turning off additional features, like backlighting, to extend battery life.

Conclusion

Baofeng radios offer preppers and survivalists a powerful, independent communication method when standard systems fail. By planning frequencies, maintaining a battery rotation, and knowing how to operate in various disaster scenarios, you equip yourself with a resilient lifeline. In the next chapter, we'll cover additional advanced skills, such as maintaining operational security and leveraging radio networks for long-term preparedness and resilience.

Chapter 11: Using Repeaters to Extend Your Range

One of the limitations of handheld radios is their relatively short range, especially in obstructed or urban environments. However, repeaters can help bridge this gap by amplifying your signal and extending your range significantly. This chapter provides an introduction to repeaters, how they work, and a step-by-step guide to finding, programming, and accessing local repeaters with your Baofeng radio.

What Are Repeaters?

Repeaters are powerful radio stations placed on elevated locations, like hilltops, tall buildings, or dedicated towers. Their purpose is to receive a signal on one frequency and instantly retransmit it on another, thereby boosting the range of the original transmission. By connecting to a repeater, you can reach users or locations well beyond the capabilities of direct, line-of-sight communication.

How Repeaters Amplify and Relay Signals

When you transmit on a specific frequency, the repeater receives your signal and simultaneously retransmits it at a stronger power level and on a separate frequency. This setup allows your signal to travel further and reach more listeners.

Repeaters typically operate with an **input frequency** (which is what your radio transmits on) and an **output frequency** (which is what your radio receives). The difference between these frequencies is called the **offset**, and it varies based on the frequency band and regional standards.

- **Offset**: This small gap between input and output frequencies ensures that your transmission doesn't interfere with the repeater's output. Most VHF repeaters in the U.S., for example, have a standard offset of ±600 kHz, while UHF repeaters generally have a ±5 MHz offset.
- **CTCSS Tones**: To prevent unauthorized or accidental access, many repeaters require a specific tone, known as **Continuous Tone-Coded Squelch System (CTCSS)** or "PL tone." By setting this tone on your radio, you can access the repeater and avoid interference from other users.

Finding and Programming Local Repeaters

Connecting to a local repeater can be a game-changer for range, but it's essential to find and correctly program it into your Baofeng radio. Here's how to get started:

Tools for Locating Repeaters

To identify repeaters in your area, several tools and databases can help.

- **RepeaterBook**: This free online database allows you to search by location, frequency, and other parameters to find nearby repeaters. RepeaterBook also offers a mobile app, making it convenient to locate repeaters on the go.
- **ARRL (American Radio Relay League) Database**: The ARRL maintains a comprehensive directory of U.S. repeaters, often available as a publication or digital file for members. This database is especially useful for those who want a reliable, periodically updated list.
- **Local Amateur Radio Clubs**: Many amateur radio clubs operate and maintain repeaters, sometimes publishing information about access and usage in newsletters or on club websites. Joining a club can also be a good way to learn local repeater etiquette and best practices.

Programming Repeater Frequencies into Your Radio

Once you have the information for a local repeater, programming it into your Baofeng radio is a straightforward process. The steps below guide you through entering the frequency, setting the offset, and configuring any required CTCSS tones.

1. **Enter the Repeater's Output Frequency**:
   o In Frequency Mode, use the keypad to enter the repeater's output frequency, which is the frequency you'll hear when the repeater transmits.
2. **Set the Offset**:
   o Access the menu and navigate to the **Offset** option, then input the offset value specified for the repeater. Ensure you enter the correct offset direction (positive or negative) based on the repeater's listing.
3. **Configure CTCSS Tone (if required)**:
   o Most repeaters require a CTCSS tone to access them. Locate the CTCSS or **Tone** setting in your menu, then enter the tone frequency associated with the repeater.
4. **Save to a Channel**:
   o Once you've entered the necessary information, save the settings to a memory channel for quick access later. This step prevents you from having to reprogram the repeater each time you want to connect.

Accessing a Repeater

After programming, you'll be ready to access the repeater. Understanding how to engage with the repeater effectively will ensure successful transmissions and allow you to make the most of its extended range capabilities.

Making Your First Call on a Repeater

When using a repeater, follow standard radio etiquette to ensure smooth communication with other users.

1. **Identify Yourself**: Start by stating your call sign and checking if the repeater is free (e.g., "This is [Your Call Sign] checking the repeater.").
2. **Wait for a Response**: After identifying yourself, wait a few seconds. Repeaters are often shared by several users, so give others a chance to respond or acknowledge you.
3. **Engage in Conversation or Pass Traffic**: If someone responds, you can begin your conversation. Keep transmissions brief and to the point to minimize occupying the repeater unnecessarily.
4. **End Your Call**: When finishing, sign off with your call sign again to let others know you are done transmitting.

Monitoring and Scanning Repeaters

Some Baofeng radios allow you to scan pre-programmed channels. This feature enables you to monitor several repeaters at once, listening for active frequencies or emergency broadcasts.

- **Scanning Mode**: Set your radio to scan mode to cycle through pre-programmed channels, stopping briefly on any active frequency. This function is useful for gathering information during emergencies or staying aware of local radio activity.

Conclusion

Repeaters are invaluable tools for extending the reach of your Baofeng radio, making it possible to communicate over distances that would otherwise be out of range. By learning how to locate, program, and use repeaters, you unlock an essential capability of your radio. In the next chapter, we'll cover advanced techniques, such as simplex communication and linking repeaters, to further enhance your radio skills and prepare you for a range of communication scenarios.

Chapter 12: Troubleshooting Common Issues

Like any device, Baofeng radios can occasionally encounter problems that impact their performance. This chapter provides a practical guide for diagnosing and resolving some of the most common issues you may experience with your Baofeng radio, from signal reception to battery maintenance and software fixes.

Poor Signal Reception

A weak or inconsistent signal is one of the most frequent challenges radio users face. Reception can be impacted by various factors, including location, environment, and equipment.

Identifying Causes of Poor Reception

1. **Location and Environment**: Dense urban areas, mountainous terrain, or even the interior of buildings can cause signal interference. Obstructions between you and the transmitting station can weaken or block your signal.
2. **Antenna Quality**: The stock antenna that comes with many Baofeng radios is functional but often lacks the strength needed for optimal performance in all environments. A low-quality or poorly attached antenna can limit range and signal clarity.

Improving Signal Reception

- **Upgrade the Antenna**: Switching to a higher-quality antenna, such as a longer whip or specialized dual-band antenna, can significantly improve range and clarity. Consider investing in a magnetic mount antenna if you plan to use the radio in a vehicle.
- **Adjust Your Position**: Moving to a higher elevation or changing locations to avoid interference (like large metal structures) can help improve reception. Positioning yourself near a window or going outdoors may also enhance your signal.
- **Use a Repeater**: If you are in an area with repeaters, programming your radio to access one can extend your reach and improve reception significantly.

Battery Problems

Keeping your radio powered and ready is essential, especially if you're relying on it for emergencies. Understanding how to properly maintain your Baofeng's battery will maximize its lifespan and performance.

Common Battery Issues

1. **Rapid Drain**: Excessive power draw from transmitting on high power or frequent use can deplete the battery more quickly than expected. Some features, like the backlight or high transmit power, can accelerate battery drain.
2. **Battery Degradation**: Over time, all rechargeable batteries degrade, losing capacity and holding less charge. If your battery isn't lasting as long as it once did, it may be nearing the end of its useful life.

Maximizing Battery Life

- **Charge Properly**: Avoid overcharging or fully depleting your battery regularly, as this can shorten its lifespan. Charging your battery when it reaches 20–30% and removing it from the charger when full will help maintain capacity.
- **Use Power-Saving Settings**: Reducing backlight duration, lowering transmit power when possible, and turning off non-essential features like dual-watch mode will help conserve battery power.
- **Invest in Backup Batteries**: For extended use or emergency scenarios, having an extra charged battery on hand is wise. A high-capacity battery pack or a portable charger can provide additional power when you need it.
- **Cold Weather Tips**: Batteries discharge faster in cold conditions. If using your radio in colder climates, keep the radio close to your body or store it in an insulated case when not in use to keep the battery warm.

Software Issues

Programming and software can occasionally present issues, especially for those who use the CHIRP programming software to customize their radio settings. Here are some of the common software-related issues and their solutions.

Common Software Problems

1. **CHIRP Connectivity Issues**: Sometimes, CHIRP may not recognize your radio, or you may experience connectivity issues during programming. This problem could stem from incompatible drivers or incorrect port settings.
2. **Incorrect Frequency Data**: Misconfigured or missing settings can lead to frequency programming errors, where the radio won't operate on saved channels as expected.
3. **Factory Reset and Configuration Loss**: Occasionally, a reset may be necessary to troubleshoot, but it can result in the loss of saved channels and settings.

Troubleshooting Software Issues

- **CHIRP Compatibility**: Ensure you're using a compatible version of CHIRP for your radio model. For Windows users, it may be necessary to install a specific driver (like the Prolific driver) for your USB programming cable. Check that the correct COM port is selected in CHIRP to establish a proper connection.
- **Factory Reset**: If your radio is unresponsive or has corrupted settings, a factory reset may restore functionality. To reset the Baofeng UV-5R, press and hold the "MENU" button while turning the radio on. Note that this will erase all programmed channels and settings, so back up important data beforehand.
- **Manual Programming**: If CHIRP isn't working, try programming directly from the radio. It may take longer, but it allows you to input critical frequencies and tones without relying on software.

Conclusion

Understanding how to troubleshoot common issues will help you get the most out of your Baofeng radio, whether you're dealing with signal reception challenges, battery performance, or software hiccups. In the next chapter, we'll discuss advanced techniques and maintenance routines to ensure your radio stays in peak working condition for the long term.

Chapter 13: Advanced Features and Customization

Once you've mastered the basics, it's time to explore the advanced features that can take your Baofeng radio experience to the next level. Baofeng radios offer several customization options, enabling you to tailor settings for privacy, multitasking, and emergency preparedness. This chapter dives into the advanced functions of CTCSS and DCS codes, Dual Watch Mode, and Emergency Alert Mode, showing you how to configure these features for optimal use.

CTCSS and DCS Codes

Continuous Tone-Coded Squelch System (CTCSS) and Digital-Coded Squelch (DCS) are coding methods that help reduce unwanted interference by filtering signals. Although sometimes referred to as "privacy codes," they do not secure communications; rather, they allow your radio to respond only to specific coded transmissions, reducing noise from other radio users on the same channel.

Understanding CTCSS and DCS

1. **CTCSS Codes**: These codes are continuous low-frequency tones transmitted along with your main transmission. Radios on the same frequency with matching CTCSS tones can hear each other, while other signals on the same frequency will be ignored, creating a form of selective listening.
2. **DCS Codes**: Digital Coded Squelch is similar but relies on a digital code rather than a continuous tone. It offers more combinations, making it a popular option in crowded radio environments.

How to Use CTCSS and DCS on Your Baofeng Radio

- **Setting Up CTCSS/DCS**: To configure CTCSS or DCS, access the menu on your radio. Scroll to the CTCSS (or "Tone" for some models) or DCS setting, and select the desired code that matches your network or group.
- **When to Use**: These codes are particularly useful in shared spaces or crowded areas, like during group events or emergencies, where multiple users may be operating on similar frequencies. By using CTCSS or DCS codes, you reduce the noise from other transmissions, allowing for clearer group communication.

Dual Watch Mode

Dual Watch Mode, also known as "dual standby," allows you to monitor two different frequencies at the same time. This feature is beneficial when you want to stay connected to two channels simultaneously, such as a primary communication channel and an emergency or backup channel.

How Dual Watch Mode Works

Dual Watch Mode alternates listening between two frequencies, allowing your radio to automatically switch to the frequency with an active signal. When one frequency receives a transmission, the radio pauses monitoring the second frequency, resuming once the transmission ends.

Setting Up and Using Dual Watch Mode

1. **Enabling Dual Watch Mode**: In the Baofeng UV-5R and similar models, go to the menu and locate the "TDR" (Dual Watch) option. Set it to "ON" to enable dual-frequency monitoring.
2. **Selecting Frequencies**: Choose your primary and secondary frequencies and enter them into the radio. These might include a local repeater for regular communication and a national or prepper frequency for emergency broadcasts.
3. **When to Use Dual Watch**: Dual Watch Mode is ideal for monitoring emergency frequencies while staying tuned to your main communication channel. For preppers, this setup lets you monitor local news or alert channels without missing important group updates.

Emergency Alert Mode

Baofeng radios include an Emergency Alert Mode that emits a loud alarm sound on the radio speaker and sends an alert tone on your current channel. This feature is designed to help you signal for assistance in emergencies when other communication methods are unavailable.

How to Activate Emergency Alert Mode

1. **Accessing the Emergency Feature**: Most Baofeng models include a dedicated "CALL" or "ALARM" button that activates Emergency Alert Mode. Pressing and holding this button triggers the alarm, making it audible to those within range on the same channel.
2. **What Happens in Emergency Mode**: Once activated, the radio emits a loud tone that can be heard by nearby users on your frequency, alerting them to your situation. It's

essential to inform your group beforehand if you plan to use this feature, so they know what the signal means.

3. **Disabling the Alert**: To turn off Emergency Mode, simply press the button again or turn off the radio. Note that using this feature unnecessarily can cause confusion, so reserve it for genuine emergencies only.

Situations for Using Emergency Alert Mode

- **Outdoor or Wilderness Activities**: If you're separated from a group or encounter a hazard, the alert mode can help others locate you.
- **Natural Disasters**: In events like earthquakes, floods, or severe weather, you may need a way to notify others when typical methods are unavailable.
- **Urban Emergencies**: In scenarios where local cell networks fail, Emergency Alert Mode can be a way to signal distress to nearby operators on your channel.

Conclusion

Exploring the advanced features of your Baofeng radio can expand your ability to communicate clearly, manage multiple channels, and signal for help if needed. Mastering these functions gives you a reliable tool for privacy, coordination, and emergency preparedness. In the following chapter, we'll cover additional customization options to further personalize your Baofeng radio experience and maximize its effectiveness.

Chapter 14: Baofeng Radios and the Future of Communication

As technology advances, so does the role of Baofeng radios in the broader landscape of modern communication. While smartphones and wireless networks dominate daily life, radio remains essential for reliable, independent, and resilient communication—especially in critical situations where traditional networks might fail. This chapter explores the ongoing relevance of radios, methods for staying current with developments, and ways to connect with the vibrant community of radio enthusiasts.

The Role of Radios in Modern Communication

With the increased reliance on digital communication, many overlook the vital role that radios continue to play. In emergencies, during natural disasters, or in remote locations, radios are often the most reliable means of staying connected.

Reliability in Critical Situations

1. **Communication Backbone in Emergencies**: When cellular and internet networks are down, radios offer a dependable alternative. They are free from the infrastructure dependencies that wired and wireless networks rely on, making them invaluable for first responders, preppers, and outdoor enthusiasts alike.
2. **Independence from Major Providers**: Radios, unlike cellular networks, do not depend on subscription services or centralized control. This makes them an ideal choice for those who prioritize self-reliance and the freedom to communicate without relying on telecommunications infrastructure.
3. **Remote Coverage**: In areas where mobile networks struggle to provide coverage, radios are often the only option for reaching others. Whether it's a remote camping trip or a wilderness expedition, Baofeng radios ensure you can communicate across rugged terrains or sparse areas.

How to Stay Updated

The world of radio, like any other field, continues to evolve. New regulations, frequency allocations, and technological advancements mean that staying informed is key to making the most of your Baofeng radio.

Keeping Up with Regulations

The Federal Communications Commission (FCC) periodically updates rules regarding permissible frequencies, license requirements, and usage protocols. It's essential to stay aware of these changes to avoid penalties and to operate legally.

1. **Monitor FCC Announcements**: The FCC website regularly posts updates on regulations and licensing information. Consider subscribing to newsletters or setting alerts to receive timely updates.
2. **Join Radio News Channels and Resources**: Websites like the American Radio Relay League (ARRL) and popular forums offer resources on evolving regulations, best practices, and technological advancements.

Exploring New Baofeng Models and Features

Baofeng consistently releases new models with improved features and enhanced performance. Staying up to date with the latest options allows you to consider upgrades or learn about added functions, like better battery life, more extensive frequency coverage, and enhanced durability.

- **Regularly Check Product Announcements**: Following Baofeng's official channels or major online marketplaces can help you stay in the loop about new models and firmware updates.
- **Explore Reviews and Comparisons**: Online reviews, particularly those on ham radio forums, provide detailed insights into how new models perform in real-world settings.

Getting Involved in the Radio Community

One of the best ways to expand your knowledge and improve your skills is by connecting with fellow radio enthusiasts. The radio community is diverse, encompassing ham operators, preppers, hobbyists, and survivalists. By joining clubs and online forums, you can benefit from a wealth of shared experience and expertise.

Joining Local and National Ham Radio Clubs

1. **Finding a Club Near You**: Local ham radio clubs are excellent for in-person learning and networking. They often host events, workshops, and emergency communication drills. Many clubs also provide mentorship to new operators, helping you navigate licensing and advanced radio techniques.
2. **National Associations**: Groups like the American Radio Relay League (ARRL) connect operators across the country, offering members exclusive resources, courses, and

certifications. National associations also advocate for radio operators' interests in regulatory and legislative matters.

Participating in Online Forums and Prepper Groups

The internet is full of communities dedicated to ham radio, emergency preparedness, and Baofeng users specifically. These platforms are perfect for asking questions, troubleshooting issues, and learning new tips from seasoned operators.

- **Radio-Specific Forums**: Sites like RadioReference, QRZ Forums, and Reddit's radio communities offer spaces where enthusiasts discuss everything from equipment reviews to advanced signal troubleshooting.
- **Prepper Networks**: Many preppers prioritize radio knowledge as part of their emergency preparedness. Joining prepper groups can provide additional insights into how to integrate radio usage into comprehensive preparedness plans, with a particular focus on using radios in worst-case scenarios.

Building Long-Lasting Connections

Networking within these communities can also lead to long-lasting friendships and a deeper appreciation of the radio field. Whether you're participating in a group event, joining a repeater network, or simply discussing tips online, staying engaged with other radio users ensures you'll continue learning and refining your skills.

Conclusion

Baofeng radios, though simple in design, are powerful tools with a lasting place in modern communication. In an age dominated by technology dependent on sprawling infrastructures, Baofeng radios remain a reliable means of direct and independent communication. By staying informed on the latest developments and connecting with others in the radio community, you can fully harness the potential of your Baofeng and ensure you're prepared for whatever comes your way. The next chapter will delve into advanced emergency communication setups, providing you with skills that take your knowledge beyond the basics.

Appendix

American Radio Frequencies

Here's a detailed list of commonly used frequencies in the United States, covering amateur, GMRS, FRS, and emergency purposes. Use these as a reference for programming your Baofeng radio, but always verify your specific region's regulations and limitations before operating on any channel.

1. **Amateur (Ham) Radio Bands**
   - **2 Meter Band (VHF)**: 144-148 MHz (Commonly used for local communications, including repeaters)
   - **70 cm Band (UHF)**: 420-450 MHz (Used for local repeaters and short-range communication)
2. **General Mobile Radio Service (GMRS)**
   - **Channels 1-7 (Shared with FRS)**: 462.5500 - 462.7250 MHz
   - **Channels 15-22**: 462.5500 - 467.7250 MHz (Requires a GMRS license)
3. **Family Radio Service (FRS)**
   - **Channels 1-22**: 462.5500 - 462.7250 MHz (FRS does not require a license and is useful for family and personal communication)
4. **Multi-Use Radio Service (MURS)**
   - **151.8200, 151.8800, 151.9400, 154.5700, 154.6000 MHz** (License-free in the U.S., suitable for personal and business use)
5. **Emergency and Prepper Frequencies**
   - **National Simplex Frequency**: 146.520 MHz (Amateur calling frequency for emergencies)
   - **Prepper Communication Networks**: Various regional frequencies available through prepper organizations
6. **NOAA Weather Radio Frequencies**
   - 162.400 MHz to 162.550 MHz (Weather updates and alerts)

Glossary of Radio Terms

This glossary provides definitions for key technical terms used throughout the book. Familiarize yourself with these terms to better understand the radio world and enhance your learning experience.

- **Antenna**: A component that transmits and receives radio waves.
- **Bandwidth**: The range of frequencies a channel can cover.
- **CHIRP**: A free programming software used for radios, including Baofeng, to facilitate easy frequency input and storage.
- **CTCSS**: Continuous Tone-Coded Squelch System, a tone sent along with a signal to reduce interference from other stations.
- **CTCSS Tone**: A low-frequency audio tone used for selective calling on a shared frequency.
- **DCS**: Digital Coded Squelch, a digital version of CTCSS that prevents interference on shared channels.
- **Dual Band**: Refers to a radio's capability to operate on both VHF and UHF frequencies.
- **Dual Watch Mode**: A feature allowing the radio to monitor two frequencies at once.
- **Frequency**: The rate at which an electromagnetic wave oscillates, measured in hertz (Hz).
- **FRS**: Family Radio Service, a set of frequencies available for general use without a license.
- **GMRS**: General Mobile Radio Service, frequencies that require a license for communication.
- **Ham Radio**: Another term for amateur radio, used by licensed operators for non-commercial purposes.
- **Modulation**: The process of varying a radio signal to transmit data.
- **NOAA**: National Oceanic and Atmospheric Administration, which operates weather radio services.
- **Offset**: The frequency difference between input and output on a repeater.
- **Repeater**: A station that extends the range of radio signals by receiving and retransmitting them.
- **Simplex**: Direct communication between two radios on the same frequency, without using a repeater.
- **Squelch**: A control that suppresses background noise when no transmission is being received.
- **VHF**: Very High Frequency, ideal for outdoor and rural communication due to its ability to cover longer distances.

- **UHF**: Ultra High Frequency, commonly used for indoor or urban settings because of its ability to penetrate obstacles.

Resource List

Here are some valuable tools, websites, and apps that will support your journey in learning and using Baofeng radios. From finding frequencies to staying updated on regulations, these resources provide a wealth of information to help you get the most out of your Baofeng radio.

1. **Websites and Databases**
    - **FCC Official Website**: fcc.gov
      Provides regulatory information, license details, and updates on legal requirements.
    - **American Radio Relay League (ARRL)**: arrl.org
      Offers courses, resources, and a membership option for licensed ham operators.
    - **RepeaterBook**: repeaterbook.com
      An online database of repeaters worldwide, essential for finding local repeaters to extend your range.
    - **RadioReference**: radioreference.com
      A comprehensive site with information on frequencies, radios, and forums for community support.
2. **Software**
    - **CHIRP**: chirp.danplanet.com
      Free, downloadable software for programming Baofeng and other radios, making frequency management simpler.
3. **Apps**
    - **NOAA Weather Radio** (App Store/Google Play)
      Access NOAA weather alerts directly from your mobile device.
    - **EchoLink** (App Store/Google Play)
      Allows licensed amateur radio operators to connect globally via VoIP technology.

- o **Zello Walkie Talkie** (App Store/Google Play)
  A popular push-to-talk app that allows radio-like communication using mobile networks.
4. **YouTube Channels**
   - o **Ham Radio Crash Course**: Explains practical radio concepts and equipment, with an emphasis on prepper-friendly content.
   - o **Signal Search**: Focuses on Baofeng radios, providing tips for new users, programming advice, and accessory recommendations.

This appendix equips you with tools, terms, and resources to fully engage with the Baofeng radio experience. Whether you're getting started or looking to expand your skills, these materials will provide ongoing support as you explore the world of radio communication.

Conclusion

In this final chapter, we'll bring together the core concepts and insights discussed throughout the book. Baofeng radios offer a versatile and powerful tool for communication, especially in emergency situations or environments where traditional networks may not be reliable. Whether you're a ham radio enthusiast, a prepper, or someone simply interested in expanding communication skills, the knowledge you've gained here is a strong foundation for confident and effective radio use.

Recap of Key Points

Final Thoughts

Your journey with Baofeng radios doesn't end here; in fact, it's just beginning. These radios open a gateway to a dynamic and skilled community of operators who share a passion for communication, preparedness, and exploration. To truly master your radio, keep practicing and experimenting with different settings, frequencies, and features.

Whether you're connecting with other operators locally or preparing for emergency scenarios, your skills and knowledge will only improve with hands-on use and continued learning. Staying engaged in ham radio clubs, online communities, or prepper networks offers new perspectives, tips, and a sense of camaraderie with like-minded individuals.

In the unpredictable world we live in, being prepared is an advantage. Baofeng radios empower you with the independence and capability to stay connected in almost any situation. Keep your radio charged, tuned, and ready—because staying connected and informed can make all the difference.

Baofeng original handbook

CONTENT

1.-SAFETY INFORMATION:

The following safety precautions shall always be observed during operation, service and repair of this equipment.

- This equipment shall be serviced by qualified technicians only.
- Do not modify the radio for any reason.
- Use only BAOFENG supplied or approved batteries and chargers.
- Do not use any portable radio that has a damaged antenna. If a damaged antenna comes into contact with your skin, a minor burn can result.
- Turn off your radio prior to entering any area with explosive and flammable materials.
- Do not charge your battery in a location with explosive and flammable materials.
- To avoid electromagnetic interference and/or compatibility conflicts, turn off your radio in any area where posted notices instruct you to do so.
- Turn off your radio before boarding an aircraft. Any use of a radio must be in accordance with airline regulations or crew instructions.
- Turn off your radio before entering a blasting area.
- For vehicles with an air bag, do not place a radio in the area over an air bag or in the air bag deployment area.
- Do not expose the radio to direct sunlight over a long time, nor place it close to heating source.
- When transmitting with a portable radio, hold the radio in a vertical position with the microphone 3 to 4 centimeters away from your lips. Keep antenna at least 2.5 centimeters away from your body when transmitting.

 WARNING: If you wear a radio on your body, ensure the radio and its antenna are at least 2.5 centimeters away from your body when transmitting.

2.-FEATURES AND FUNCTIONS:

- Dual-band handheld transceiver with display function menu on the display "LCD".
- DTMF encoded.
- Lithium-ion battery with high capacity.
- Commercial FM radio receiver (65 MHz ~ 108 MHz).
- Incorporates 105 codes "DCS" and 50 privacy codes "CTCSS" programmable.
- Function "VOX" (voice operated transmission).
- Alarm function.
- Up to 128 memory channels.
- Broadband (Wide) / Narrowband (Narrow), selectable.
- High power / low (5 W/1 W) selectable.
- Display illumination and programmable keyboard.
- Function "beep" on the keyboard.
- Dual Watch/dual reception .
- Selectable Frequency Step 2.5/5/6.25/10/12.5/25 kHz.
- Function "OFFSET" (frequency offset for repeater access).
- Battery saving function "SAVE".
- Timer transmission "TOT" programmable.
- Selecting the Scan Mode.
- Function Busy Channel Lock "BCLO".
- Built-in RX CTCSS/DCS scan
- Built-in LED flashlight.
- Programmable by PC.

- Level Threshold "Squelch" adjustable from 0 to 9.
- Crossband reception
- Tone end of transmission
- Built-in key lock

3.-UNPACKING AND CHECKING EQUIPMENTS:

Carefully unpack the transceiver. We recommend that you identify the items listed in the following before discarding the packing material. If any items are missing or have been damaged during shipment, please contact your dealers immediately.

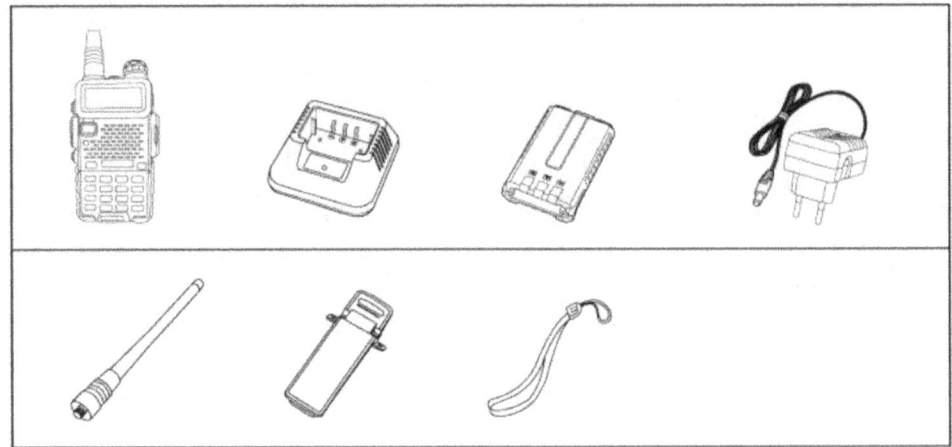

Note:
- Items included in the package, may differ from those listed in the table above depending on the country of purchase. For more information, consult your dealer or vendor.

4.- OPTIONAL ACCESSORIES:

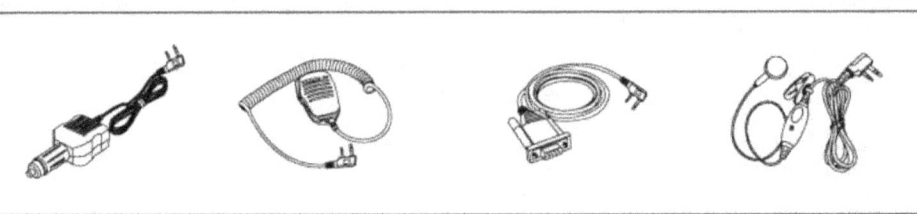

Note:
- Consult the dealer or retailer for information about options available.

04

5.- INSTALLATION OF ACCESSORIES:

5.1.- INSTALLING THE ANTENNA:

Install the antenna as shown in the figure below and turn it clockwise until it stops.

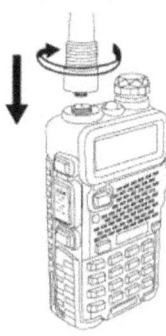

Note:
- When installing the antenna, don't rotate it by its top, holding it by its base and turn.
- If you use an external antenna, make sure the 'SWR' is about 1.5:1 or less, to avoid damage to the transceiver's final transistors.
- Do not hold the antenna with your hand or wrap the outside of it to avoid bad operation of the transceiver.
- Never transmit without an antenna.

5.2.- INSTALLING THE BELT CLIP:

If necessary, install the belt clip at the rear of the battery compartment cover as shown in the figure below.

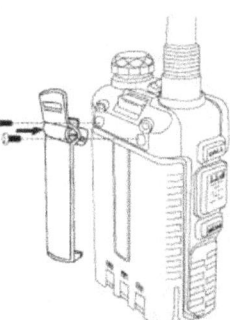

Note:
- Do not use any kind of glue to fix the screw on the belt clip. The solvents Glue may damage the battery casing.

5.3.- MICRO-HEADSET INSTALLATION OF EXTERNAL:

Plug the external micro-headset connector into the jack of 'SP. & MIC' of the transceiver as shown in the figure below.

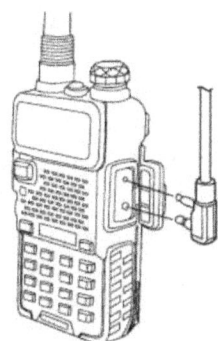

05

5.4.- BATTERY INSTALLATION:

-When attaching the battery, make sure the battery is in parallel and in good contact with the aluminum chassis. The battery bottom is about 1 to 2 centimeters below the bottom of the radio's body.
-Align the battery with the guide rails on the aluminum chassis and slide it upwards until a 'click' is heard.
-The battery latch at the bottom locks the battery.

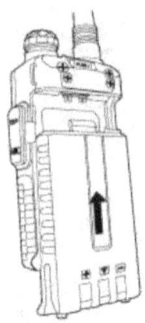

-Turn off the radio before removing the battery.
-Slide the battery latch, at the bottom of the radio's body, in the direction indicated by the arrow.
-Slide down the battery for about 1 to 2 centimeters, and then remove the battery from the radio's body.

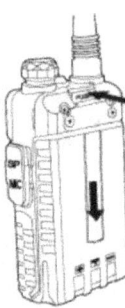

6.-BATTERY CHARGING:

Use only the charger specified by the manufacturer. The charger's LED indicates the charging progress.

CHARGING STATUS	LED INDICATION
Standby (no-load)	Red LED flashes,while Green LED glows
Charging	Red LED solidly glows
Fully Charged	Green LED solidly glows
Error	Red LED flashes,while Green LED glows

06

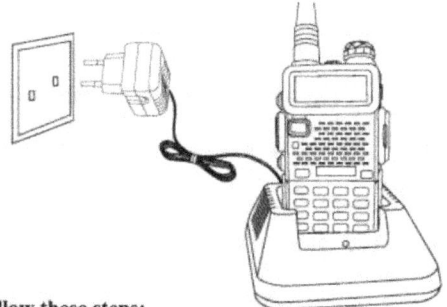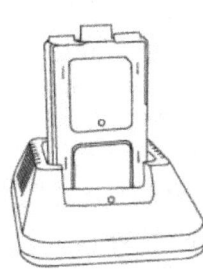

Please follow these steps:
1. Plug the power cord into the adapter.
2. Plug the AC connector of the adapter into the AC outlet socket.
3. Plug the DC connector of the adapter into the DC socket on the back of the charger.
4. Place the radio with the battery attached, or the battery alone, in the charger.
5. Make sure the battery is in good contact with the charging terminals. The charging process initiates when the red LED lights.
6. The green LED lights about 4 hours later indicating the battery is fully charged. Then remove the radio with the battery attached or the battery alone from the charger.

7. -BATTERY INFORMATION:

7.1.-INITIAL USE

New batteries are shipped uncharged fully from the factory. Charge a new battery for 5 hours before initial use. The maximum battery capacity and performance is achieved after three full charge/discharge cycles. If you notice the battery power runs low, please recharge the battery.

 WARNING: -To reduce the risk of injury, charge only the battery specified by the manufacturer. Other batteries may burst, causing bodily injury and property damage. -To avoid risk of personal injury, do not dispose of batteries in a fire!

-Dispose of batteries according to local regulations (e.g. recycling). Do not dispose as household waste.
-Never attempt to disassemble the battery.

7.2.-BATTERY TIPS:

1. When charging your battery, keep it at a temperature among 5℃ - 40℃. Temperature out of the limit may cause battery leakage or damage.
2. When charging a battery attached to a radio, turn the radio off to ensure a full charge.
3. Do not cut off the power supply or remove the battery when charging a battery.
4. Never charge a battery that is wet. Please dry it with a soft cloth prior to charge.
5. The battery will eventually wear out. When the operating time (talk-time and standby time) is noticeably shorter than normal performance, it is time to buy a new battery.

7.3.-PROLONG BATTERY LIFE:

1. Battery performance will be greatly decreased at a temperature below 0℃. A spare battery is necessary in cold weather. The cold battery unable to work in this situation may work under room temperature, so keep it for later use.

07

2. The dust on the battery contact may cause the battery cannot work or charge. Please use a clean dry cloth to wipe it before attaching the battery to the radio.

7.4.-BATTERY STORAGE:

1. Fully charge a battery before you store it for a long time, to avoid battery damage due to over-discharge.
2. Recharge a battery after several months' storage (Li-Ion batteries: 6 months), to avoid battery capacity reduction due to over-discharge.
3. Store your battery in a cool and dry place under room temperature, to reduce self-discharge.

8.-PARTS, CONTROLS AND KEYS:

8.1.-RADIO OVERVIEW:

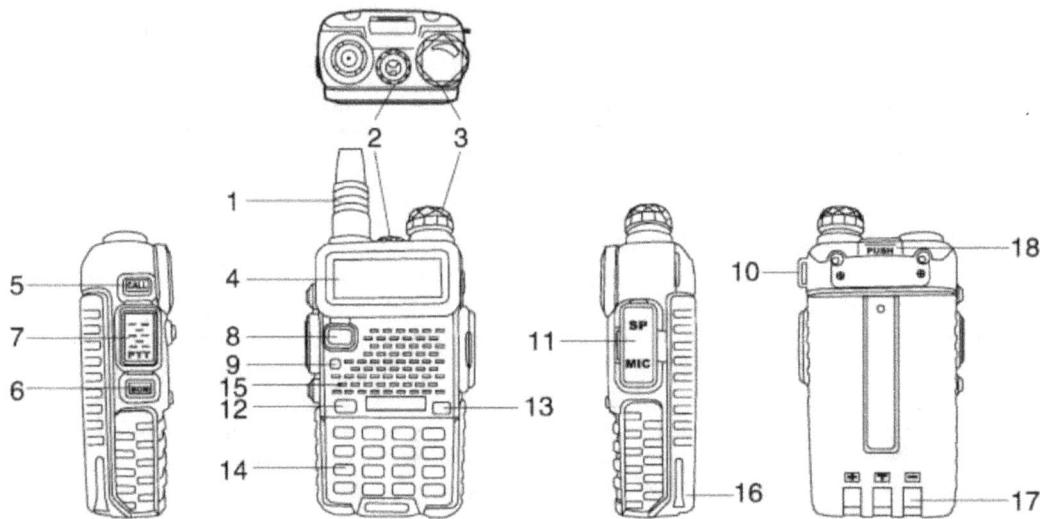

1. antenna
2. flashlight
3. knob (ON/OFF,volume)
4. LCD
5. SK-side key1/CALL(radio,alarm)
6. SK-side key2/MONI(flashlight,monitor)
7. PTT key(push-to-talk)
8. VFO/MR (frequency mode/channel mode)
9. LED indicator

10. strap buckle
11. accessory jack
12. A/B key(frequency display switches)
13. BAND key(band switches)
14. keypad
15. SP.&MIC.
16. battery pack
17. battery contacts
18. battery remove button

08

8.2.- COMMAND/KEY DEFINITION:

▶ [PTT](PUSH-TO-TALK):
Press and hold down the [PTT] button to transmit; release it to receive.

▶ SK-SIDE KEY1/[CALL]:
- Press the [CALL] button,to activate the FM Radio;Press it again to deactivate the FM Radio.
- Press and hold on the [CALL]button,to activate the alarm function; Press and hold it again,to deactivate the alarm function.

▶ SK-SIDE KEY2/[MONI]:
-Press the [MONI] button,to turn on the flashlight;Press it again to turn off.Press and hold on the [MONI] button,to monitor the signal.

▶[VFO/MR]BUTTON:
-Press the [VFO/MR] button,to switch the frequency mode and channel mode.

▶[A/B]BUTTON:
-Press the [A/B] button,to switch frequency display.

▶[BAND]BUTTON:
-Press the [BAND]button,to switch band dispaly.
-While FM radio being activated,press the [BAND]button to switch the band of FM radio(band 65-75MHz/76-108MHz).

▶[★SCAN]KEY:
-Press the [★SCAN] key to activate the Reverse function,it will exchange a separate reception and transmission frequency.
-Press the [★SCAN] key for 2 seconds to start scanning(frequency/channel).
- While FM radio being activated,press the [★SCAN] key to search FM radio station.
-While setting the RX CTCSS/DCS, press the key [★SCAN] to scan the RX CTCSS/DCS.

▶[#🔒] KEY:
-Under channel mode, press [#🔒] key to switch High/Low transmit power.
-Press [#🔒] key for 2 seconds to lock/unlock the keypad.

▶FUNCTION KEYPAD:
-[MENU]key:
-To enter the menu of the radio and confirm the setting.
-[▲][▲]key:
-Press and hold [▲]or[▲]key for frequency up or down fast.
-Press [▲]or[▲]key,the scanning will be opposite.
-[EXIT]key:
-To cancel /clear or exit.

▶NUMERIC KEYPAD:
-Used to enter information for programming the radio's lists and the non-standard CTCSS
-Under transmission mode, press the numeric key to send the signal code(the code should be set by PC software).

▶ ACCESSORY JACK:
-The jack is used to connect audio accessories, or other accessories such as programming cable.

9.-'LCD' DISPLAY:

The display icons appear when certain operations or specific features are activated.

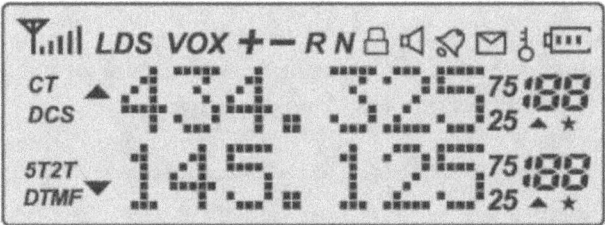

Icon	Description
188	Operating channel.
75 25	Operating frequency.
CT	'CTCSS' activated.
DCS	'DCS' activated.
+−	Frequency offset direction for accessing repeaters.
S	Dual Watch/Dual Reception functions activated.
VOX	Function 'VOX' enabled.
R	Reverse function activated.
N	Wide Band selected.
🔋	Battery Level indicator
🔑	Keypad lock function activated.
L	Low transmit power.
▲▼	Operation frequency.
📶	Signal Strength Level.

10

10.- 1750 Hz TONE FOR ACCESS TO REPEATERS:

The user needs to establish long distance communications through an amateur radio repeater which is activated after receiving a 1750 Hz tone. Press and hold on the [PTT],then press the [BAND] button to transmit a 1750Hz tone.

11.- BASIC OPERATION:

11.1.-RADIO ON-OFF/VOLUME CONTROL :
-Make sure the antenna and battery are installed correctly and the battery charged.
-Rotate the knob clockwise to turn the radio on, and rotate the knob fully counter-clockwise until a 'click' is heard to turn the radio off. Turn the knob clockwise to increase the volume, or counter-clockwise to decrease the volume.

11.2.- SELECTING A FREQUENCY OR CHANNEL:
-Press the key[▲]or[▲]to select the desired frequency/channel you want. The display shows the frequency / channel selected.
-Press and hold down the key [▲]or[▲] for frequency up or down fast.

Note:
- You can not select a channel if not previously stored.

12-ADVANCED OPERATION:

You can program your transceiver operating in the setup menu to suit your needs or preferences.

12.1.-SET MENU DESCRIPTION:

Menu	Function/Description	Available settings
0	SQL (Squelch level)	0-9
1	STEP(Frequency step)	2.5/5/6.25/10/12.5/25kHz
2	TXP(Transmit power)	HIGH/LOW
3	SAVE(Battery save,1:1/1:2/1:3/1:4)	OFF/1/2/3/4
4	VOX(Voice operated transmission)	OFF/0-10
5	W/N(Wideband/narrowband)	WIDE/NARR
6	ABR(Display illumination)	OFF/1/2/3/4/5s
7	TDR(Dual watch/dual reception)	OFF/ON

11

8	BEEP(Keypad beep)	OFF/ON
9	TOT(Transmission timer)	15/30/45/60.../585/600seconds
10	R-DCS(Reception digital coded squelch)	OFF/D023N...D754I
11	R-CTS(Reception Continuous Tone Coded Squelch)	67.0Hz...254.1Hz
12	T-DCS(Transmission digital coded squelch)	OFF/D023N...D754I
13	T-CTS(Transmission Continuous Tone Coded Squelch)	67.0Hz...254.1Hz
14	VOICE(Voice prompt)	OFF/ON
15	ANI(Automatic number identification of the radio,only can be set by PC software.	
16	DTMFST(The DTMF tone of transmitting code.)	OFF/DT-ST/ANI-ST/DT+ANI
17	S-CODE(Signal code, only could be set by PC software.)	1,...,15 groups
18	SC-REV(Scan resume method)	TO/CO/SE
19	PTT-ID(press or release the PTT button to transmit the signal code)	OFF/BOT/EOT/BOTH
20	PTT-LT(delay the signal code sending)	0,...,30ms
21	MDF-A(under channel mode, A channel displays. Note: name display only can be set by PC software.	FREQ/CH/NAME
22	MDF-B(under channel mode, B channel displays. Note: name display only can be set by PC software.	FREQ/CH/NAME
23	BCL(busy channel lockout)	OFF/ON
24	AUTOLK(keypad locked automatically)	OFF/ON
25	SFT-D(direction of frequency shift)	OFF/+/-
26	OFFSET(frequency shift)	00.000...69.990
27	MEMCH(stored in memory channels)	000, ...127
28	DELCH(delete the memory channels)	000, ...127
29	WT-LED(illumination display color of standby)	OFF/BLUE/ORANGE/PURPLE
30	RX-LED(illumination display color of reception)	OFF/BLUE/ORANGE/PURPLE
31	TX-LED(illumination display color of transmitting)	OFF/BLUE/ORANGE/PURPLE
32	AL-MOD(alarm mode)	SITE/TONE/CODE
33	BAND(band selection)	VHF/UHF
34	TX-AB(transmitting selection while in dual watch/ reception)	OFF/A/B
35	STE(Tail Tone Elimination)	OFF/ON

12

36	RP_STE(Tail tone elimination in communication through repeater)	OFF/1,2,3...10
37	RPT_RL(Delay the tail tone of repeater)	OFF/1,2,3...10
38	PONMGS(Boot display)	FULL/MGS
39	ROGER(tone end of transmission)	ON/OFF
40	RESET (Restore to default setting)	VFO/ALL

12.2.-SHORTCUT MENU OPERATION:

1.-Press the key MENU,then press the key ▲ or ▼ to select the desired menu.
2.-Press the key MENU again, come to the parameter setting.
3.-Press the key ▲ or ▼ to select the desired parameter.
4.-Press the key MENU to confirm and save, press the key EXIT to cancel setting or clear the input.

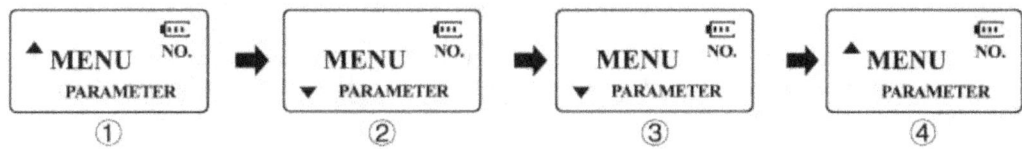

-Note:
Under channel mode,the following menu settings are invalid:CTCSS,DCS,W/N,PTT-ID,BCL,SCAN ADD TO,S-CODE,CHANNEL NAME.Only the H/L power could be changed.

12.3.-"SQL" (SQUELCH):

-The squelch mute the speaker of the transceiver in the absence of reception. With the squelch level correctly set, you will hear sound only while actually receiving signals and significantly reduces battery current consumption. It is recommended that you set Level 5.

12.4.- FUNCTION "VOX" (VOICE OPERATED TRANSMISSION):

-This function is not necessary to push the [PTT] on the transceiver for a transmission. Transmission is activated automatically by detecting the radio voice. When finish speaking, the transmission automatically terminated and the transceiver will automatically receive signal. Be sure to adjust the VOX Gain level to an appropriate sensitivity to allow smooth transmission.

12.5.- SELECT WIDEBAND OR NARROW BAND "W/N":

In areas where the RF signals are saturated, you must use the narrow band of transmission to avoid interference in adjacent channels.

12.6.- TDR (DUAL WATCH/DUAL RECEPTION):

This feature allows you to operate between frequency A and frequency B. Periodically, the transceiver checks whether a signal is received on another frequency that we have scheduled. If you receive a signal, the unit will remain in the frequency until the received signal disappears.

13

12.7.- TOT(TRANSMISSION TIMER):

This function can automatically control the time we transmit each time you press [PTT] on the transceiver. This feature is very useful to avoid overheating excessive power transistors of the transceiver. The transceiver will be off transmission automatically once the set time.

12.8.-CTCSS/DCS:

In some cases only want to establish communications in a closed user group at a particular frequency or channel, for it will use "CTCSS" or code "DCS" for reception.
The "squelch" opens only when receiving a frequency with "CTCSS" or codes "DCS" same as the programmed in your transceiver. If codes of the received signal differs from those programmed in your transceiver, the "squelch" will not open and the received signal can be heard.

Note:
- The use of "CTCSS" or "DCS" in a communication, does not guarantee complete confidentiality communication.

12.9.- ANI

-ANI (Automatic Number Identification) is also known as PTT ID because an ID is transmitted when the PTT button of the radio is pressed and/or released. This ID tells the dispatcher which field radio was keyed.
Only could be set by PC software.

12.10.- DTMFST (DTMF TONE OF TRANSMITTING CODE):

First you should set the PTT-ID as BOT/EOT/BOTH
-"OFF"-Under transmitting mode, you can't hear the DTMF tone, while you press the key to transmit the code or code automatically transmitted.
-"DT-ST"-Under transmitting mode, you can hear the DTMF tone, while you press the key to transmit the code.
-"ANI-ST"-under transmitting mode, you can hear the DTMF tone, while the code automatically transmitted.
-"DT-ANI"-under transmitting mode, you can hear the DTMF tone, while you press the key to transmit the code or the code automatically transmitted.

12.11.- SC-REV(SCAN RESUME METHOD):

This transceiver allows you to scan memory channels, all the bands or part of the bands.
When the transceiver detects a communication, the scan will stop automatically.

Notes:
- "TO" (Time Operation):
Scanning will stop when it detects an active signal. The scanning will stop on each channel or active frequency for a predetermined time, after that time the scan will resume automatically.
- "CO" (Carrier Operation):
The scanning will stop and remain in the frequency or channel, until the active signal disappears.
- "SE"(Search Operation):
The scanning will stop and remain in the frequency or channel after it detects an active signal.

14

12.12.- PTT-ID(PTT OR RELEASE PTT TO TRANSMIT THE SIGNAL CODE):

-This feature allows you to know who call you.
-"OFF"-Don't transmit the code while push the PTT button.
-"BOT"-Transmit the code while push the PTT button.(the code only could be set by PC software.)
-"EOT"-Transmit the code while release the PTT button.
-"BOTH"-Transmit the code while push or release the PTT button.

12.13.- BCL(BUSY CHANNEL LOCKOUT):

The BCLO feature prevents the radio's transmitter from being activated if a signal strong enough to break through the "noise" squelch is present. On a frequency where stations using different CTCSS or DCS codes may be active, BCLO prevents you from disrupting their communications accidentally (because your radio may be muted by its own tone decoder).

12.14.- SFT-D(DIRECTION OF FREQUENCY SHIFT):

The "OFFSET" is the difference or offset between the reception frequency and the frequency of transmission for access to amateur radio repeaters. Set the "OFFSET" according to the "OFFSET" amateur radio repeater through which want to communicate.

12.15.- OFFSET(FREQUENCY SHIFT):

When communicating via a repeater, the direction of displacement of frequency should be timed to the displacement of the transmission frequency is higher or lower than the receiving frequency.
example:
If we want to make a communication through amateur radio repeater whose frequency input is 145,000 MHz and 145,600 MHz is output, we select the "OFFSET" of the previous section in 0600 and the direction of travel "SHIFT" programmed to [-], so the transceiver will always 145,600 MHz in frequency and when you press [PTT] to transmit transceiver, the frequency will automatically move to 145,000 MHz

12.16.-STE (TAIL TONE ELIMATION):

This function is used to activate or deactivate the transmission end of the transceiver. this final tone transmission only be used in communications between transceivers and not in communications through a repeater, which must be deactivated.

13.-CTCSS TABLE:

N°	Tone(Hz)	N°	Tone(Hz)	N°	Tone(Hz)	N°	Tone(Hz)	N°	Tone(Hz)
1	67.0	11	94.8	21	131.8	31	171.3	41	203.5
2	69.3	12	97.4	22	136.5	32	173.8	42	206.5
3	71.9	13	100.0	23	141.3	33	177.3	43	210.7
4	74.4	14	103.5	24	146.2	34	179.9	44	218.1
5	77.0	15	107.2	25	151.4	35	183.5	45	225.7
6	79.7	16	110.9	26	156.7	36	186.2	46	229.1
7	82.5	17	114.8	27	159.8	37	189.9	47	233.6
8	85.4	18	118.8	28	162.2	38	192.8	48	241.8
9	88.5	19	123.0	29	165.5	39	196.6	49	250.3
10	91.5	20	127.3	30	167.9	40	199.5	50	254.1

15

14.-DCS TABLE:

N°	Code	N°	Code	N°	Code	N°	Code	N°	Code
1	D023N	22	D131N	43	D251N	64	D371N	85	D532N
2	D025N	23	D132N	44	D252N	65	D411N	86	D546N
3	D026N	24	D134N	45	D255N	66	D412N	87	D565N
4	D031N	25	D143N	46	D261N	67	D413N	88	D606N
5	D032N	26	D145N	47	D263N	68	D423N	89	D612N
6	D036N	27	D152N	48	D265N	69	D431N	90	D624N
7	D043N	28	D155N	49	D266N	70	D432N	91	D627N
8	D047N	29	D156N	50	D271N	71	D445N	92	D631N
9	D051N	30	D162N	51	D274N	72	D446N	93	D632N
10	D053N	31	D165N	52	D306N	73	D452N	94	D645N
11	D054N	32	D172N	53	D311N	74	D454N	95	D654N
12	D065N	33	D174N	54	D315N	75	D455N	96	D662N
13	D071N	34	D205N	55	D325N	76	D462N	97	D664N
14	D072N	35	D212N	56	D331N	77	D464N	98	D703N
15	D073N	36	D223N	57	D332N	78	D465N	99	D712N
16	D074N	37	D225N	58	D343N	79	D466N	100	D723N
17	D114N	38	D226N	59	D346N	80	D503N	101	D731N
18	D115N	39	D243N	60	D351N	81	D506N	102	D732N
19	D116N	40	D244N	61	D356N	82	D516N	103	D734N
20	D122N	41	D245N	62	D364N	83	D523N	104	D743N
21	D125N	42	D246N	63	D365N	84	D526N	105	D754N

15.-TECHNICAL SPECIFICATION:

15.1.-GENERAL:

Frequency range	65MHz-108MHz(Only commercial FM radio reception) VHF:136MHz-174MHz (Rx/Tx).UHF:400MHz-480MHz (Rx/Tx)
Memory channels	Up to 128 channels
Frequency stability	2.5ppm
Frequency step	2.5kHz/5kHz/6.25kHz/10kHz/12.5kHz/25kHz
Antenna impedance	50 Ω
Operating temperature	-20 ℃ to +60 ℃
Supply voltage	Rechargeable Lithium-Ion mAh 7.4V/1800
Consumption in standby	≤75mA
Consumption in reception	380mA
Consumption in transmission	≤1.4 A
Mode of operation	Simplex or semi-duplex
Duty cycle	03/03/54 min. (Rx / Tx / Standby)
Dimensions	58mm x 110mm x 32mm
Weight	130 g (approximate)

15.2. - TRANSMITTER:

RF power	4W/1W
Type of modulation	FM
Emission class	16K Φ F3E/11K Φ F3E (W/N)
Maximum deviation	≤ ± 5 kHz/≤ ± 2.5 kHz (W/N)
Spurious emissions	<−60 dB

15.3. - RECEIVER:

Receiver sensitivity	0.2 μ V(at 12 dB SINAD)
Intermodulation	60 dB
Audio output	1000mW
Adjacent channel selectivity	65/60dB

Note:
- All specifications shown are subject to change without notice.

16.-TROUBLESHOOTING:

Problem	Possible cause / solution
The radio does not start.	The battery is low, replace the battery with a charged battery or proceed to the battery. The battery is not installed correctly, remove the battery and reattach it.
The battery runs down quickly.	The battery life has come to an end, replace the battery with a new one. The battery is fully charged, make sure the battery is made in full.
The receiving indicator LED lights but do not hear the speaker.	Make sure the volume setting is too low. Make sure the undertones "CTCSS" or code "DCS" are the same as those programmed in the transceiver of the other members of your group.
When transmitting, the other members of his group do not receive the communication.	Make sure the undertones "CTCSS" or code "DCS" programmed in your transceiver are the same as those programmed in the transceiver of the other members of your group. Your partner or you, are too far. You or your partner are in a bad area of RF signal propagation.
In"standby"mode, the transceiver transmits without pressing the "PTT".	Check the level adjustment function "VOX" is not set too sensitive.
Receive communications from other user groups while communicating with your group.	Change frequency or channel. Change the undertones "CTCSS" or code "DCS" in your group.
Communication with other members of your group is poor or low quality.	You or your partner is too far away or in an area of poor radio signal propagation, such as inside a tunnel, inside an underground car park, in a mountainous area, including large metal structures, etc..
Once these checks, if you still have problems with the transceiver, check with your distributor, dealer or service center.	

17.-WARRANTY: (Better buy the radios from local dealer).

<table>
<tr><td colspan="3" align="center">WARRANTY CERTIFICATE</td></tr>
<tr><td>Brand:</td><td>Model no.:</td><td>Serial no.:</td></tr>
<tr><td colspan="2">Name of purchaser:</td><td rowspan="6"></td></tr>
<tr><td colspan="2">Address:</td></tr>
<tr><td>City:</td><td>Zip code:</td></tr>
<tr><td>Province/State:</td><td>Tel no.:</td></tr>
<tr><td colspan="2">Date of purchase:</td></tr>
<tr><td colspan="2">WARNING: Warranty is valid provided it is complete and properly filled in legibly and clearly present the seal and name of the dealer and have attached the bill proof of purchase of equipment.</td></tr>
</table>

The device described in this Certificate is guaranteed for a period of one year from the date of sale to the final user. This Warranty Certificate is unique and not transferable and may not be reissued for new or original or copy. Substitution of product failure or any part thereof shall not involve extension of the guarantee.

The warranty covers the replacement and free replacement of all parts that are defective in materials and components used in manufacturing and / or assembly of the apparatus.

The warranty does not cover any faults caused by accident, improper installation and use, electric shock (eg storms), connect a power other than that specified, reverse polarity in the diet, or claims due to deterioration in the external appearance of normal use, nor the amount or condition of the accessories. Checking the accessories is the responsibility of the purchaser at the time of purchasing the device.

The warranty does not cover rechargeable batteries even if they are part of the equipment purchased as they are considered consumables, the impairment must be reported within a period of fifteen days from the date of purchase.

The warranty is void on the following assumptions:

1. - Devices that have been manipulated by another or by anyone other than authorized service provider.
2. - Equipment and accessories in which the serial number has been altered, deleted or filed unreadable.
3. - Use of the product than as intended.

To make use of the guarantee is necessary to give the dealer or any of the Authorised Service the defective device with its accessories and the following documentation:

1. - Warranty Certificate duly completed and sealed.
2. - Original invoice which clearly identifies the device and the date of purchase.
3. - Description of the faults.

The warranty terms contained in this Certificate of Guarantee do not exclude, modify or restrict the statutory rights of the buyer by virtue of the laws in force at the time of purchase, but are added to them.

19

www.ingramcontent.com/pod-product-compliance
Lightning Source LLC
Chambersburg PA
CBHW082115220526
45472CB00009B/2190